BANKRUPTING PHYSICS

BANKRUPTING PHYSICS

HOW TODAY'S TOP SCIENTISTS ARE GAMBLING AWAY THEIR CREDIBILITY

ALEXANDER UNZICKER

AND

SHEILLA JONES

palgrave
macmillan

BANKRUPTING PHYSICS
Copyright © Alexander Unzicker and Sheilla Jones, 2013.

Translation from the German language edition: *Vom Urknall zum Durchknall*
by A. Unzicker. Copyright © Springer-Verlag Berlin Heidelberg 2010.
Springer-Verlag is part of Springer Science+Business Media.

First published in 2013 by PALGRAVE MACMILLAN® in the United States—
a division of St. Martin's Press LLC, 175 Fifth Avenue, New York, NY 10010.

Where this book is distributed in the UK, Europe and the rest of the world,
this is by Palgrave Macmillan, a division of Macmillan Publishers Limited,
registered in England, company number 785998, of Houndmills,
Basingstoke, Hampshire RG21 6XS.

Palgrave Macmillan is the global academic imprint of the above companies
and has companies and representatives throughout the world.

Palgrave® and Macmillan® are registered trademarks in the United States,
the United Kingdom, Europe and other countries.

ISBN 978–1–137–27823–4

Library of Congress Cataloging-in-Publication Data

Unzicker, Alexander.
     Bankrupting physics : how today's top scientists are gambling away
their credibility / Alexander Unzicker and Sheilla Jones.
        pages cm
     ISBN 978–1–137–27823–4 (hardback)
        1. Physics—Philosophy. 2. Physics—Experiments.
     3. Physics—Mathematical models. I. Jones, Sheilla. II. Title.
QC6.U59 2013
530—dc23   2013004524

A catalogue record of the book is available from the British Library.

Design by Newgen Imaging Systems, Ltd., Chennai, India

First edition: July 2013

10 9 8 7 6 5 4 3 2 1

Printed in the United States of America.

CONTENTS

Prologue
Relaxing, Exciting, and Overstrung Physics   ix

Part I
Shortcut

Chapter 1. Not too Bad, *Homo sapiens*, But...
Reasons for Doubt: Something Is Rotten in the State of Physics   3

Chapter 2. Galileo Would Freak Out!
A Quantum Leap in Measuring Devices: Why We
Live in Fantastic Times   19

Chapter 3. A Speedy Revolution
Why Cosmology Is Going the Wrong Way   31

Part II
Crossroads

Chapter 4. The Basic Story
What Einstein Told Us about Gravity and Space-Time   45

Chapter 5. Still a Mystery
Newton's Gravitational Constant: From England to the
Edge of the Universe   59

Chapter 6. The Riddle of Small Accelerations
Are Galaxies Really Just Big Planetary Systems?   69

Chapter 7. Lost in the Dark
Dark Matter and Dark Energy: Invisible or All in Your Mind?   81

Chapter 8. Precision in the Tea Leaves
Message from the Cosmic Microwave Background:
How Much Is Just Noise? 95

Part III
Dead End

Chapter 9. Muddy Water
The Cosmology of Dark Pixels in the First Dark Age:
Giving Work to Supercomputers 111

Chapter 10. Speculation Bubbles Rise
Expansion, Imagination, Inflation: Do We Know
There Was a First Second? 121

Chapter 11. Blacking Out
Black Holes, the Big Bang, and Quantum Gravity:
Ecological Niches for Theorists 131

Chapter 12. The Fiancée You Won't Marry
The Standard Model of Particle Physics: How Playing with
Mathematical Beauty Took Over Real Life 141

Chapter 13. Chronicle of a Surprise Foretold
How Higgsteria Delayed the Bankruptcy of Particle Physics 151

Chapter 14. New Dimensions in Nonsense
Branes, Multiverses, and Other Supersicknesses:
Physics Goes Nuts 159

Chapter 15. Goodbye Science, Hello Religion
String Theory: How the Elite Became a Sect and a Mafia 173

Part IV
Backing Up

Chapter 16. Clear Water
Reason vs. Circular Logic: How Science Should Work 191

Chapter 17. Welcome to Byzantium
Complications on Complications:
How Physics Became a Junk Drawer 203

Chapter 18. The First Wrong Turn
Deviation Decades Ago: Calculating Replaces Thinking 217

Chapter 19. The Math Fallout
How Theoretical Fashions Impede Reflection   229

Chapter 20. Big Science, Big Money, Big Bubbles
What's Wrong with the Physics Business   239

Chapter 21. Outlook
Get Prepared for the Crash   251

Appendix   255

Thanks   257

Permissions   259

Notes   261

Literature   267

Index   269

PROLOGUE

RELAXING, EXCITING, AND OVERSTRUNG PHYSICS

There is nothing more relaxing than physics. However distressing everyday problems may appear to us, our human endeavors appear trivial and inconsequential under the universe's vast and starry sky. Or, alternatively, imagine the smallness of the elementary building blocks that comprise Nature. These thoughts are a refuge for the harried mind.

But it is equally true that there is hardly anything more exciting than physics. Physics shows us the microscopic world with incredible tools, and the entire universe has opened up before our eyes with the kind of detail unknown to previous generations. Never before have we been better prepared to discover the fundamental laws of Nature.

Yet, and regrettably, there is also much that is ludicrous being sold under the name of physics, and this book will deal extensively with this issue. This is not a polemic against science. Far from it. Rather, it is a recognition that while much of our civilization is based on physics, it is human curiosity and aspiration that drives science. And inevitably, there is also human wishful thinking and human failure.

Applied physics has been a terrific success to date, and the fundamental findings of theoretical physics in the early twentieth century were among the greatest accomplishments of humankind. But that was then. Today, the major part of theoretical physics has instead gotten lost in bizarre constructs that are completely disconnected from reality, in a mockery of the methods that grounded the success of physics for 400 years.

Fortunately, an increasing number of people in the scientific community and in the public are becoming aware that bold ideas such as string theory, multiverses, "chaotic inflation," and a "holographic principle" have little to do with real physics. Unfortunately, before these fantasies took over, physics was already ailing.

A particularly worrying symptom of the current state of affairs in physics is the so-called discovery of the Higgs boson at CERN. The media-hyped announcement in 2012 has been followed up by a series of announcements, each installment making the case that the big sensation is "increasingly more likely." But what was actually discovered were a number of unexplained signals obtained by extensive filtering methods, raising many questions for everyone who takes a sober perspective. Nobody can claim to oversee the analysis of the massive pile of data produced by CERN's collider experiments.

Nevertheless, these signals are pushed to serve as evidence for the long-theorized Higgs boson supporting the "standard model" of particle physics, although this standard model is not even a well-defined theory. Such an interpretation speaks more of desperation to validate the past six decades of research and to shore up a model that is wobbling precariously under the weight of all the bits and pieces glued onto it to make it work.

I can hear the protests of physicists who conveyed the message in good faith. But they had little choice. The CERN particle search is the most expensive experiment ever conducted, and the thousands of scientists doing high-energy research there had to celebrate *any* outcome as a breakthrough, if only to justify the billions of dollars of public money being spent.

In economics, we have seen expert analysts parroting one another's euphoric statements when the underlying construction, based entirely on confidence, was about to crash. A worrying amount of knowledge in physics is based on such confidence in expert opinions.

Economics can produce bubbles, and so can science. It appears to be a universal mechanism of human aspiration that, while following the seemingly obvious, methods can gradually slip into absurdity, leaving behind unresolved problems that escape the attention of busy researchers. We need to take a step back from the scientific convictions of our generation and scrutinize the complicated models that might well be rather mundane constructions than true laws of Nature.

This book is not going to be the ally of religion, intelligent design garbage, or other esoteric hokum. It is true science that we need, and it is the erosion of its methods that has led to the crisis of physics we observe today. Keep this diagnosis in mind while you are reading the book.

Whatever my criticisms of science, I still have more esteem for scientists' self-indulgent fantasies about imaginary multiverses than for those who, in pursuit of power, wealth, and military vengeance at the expense of the environment and the well-being of humankind, are about to destroy our real world and thus put into danger the whole enterprise of *Homo sapiens* on this planet. More than ever, the world needs the competence of physicists. And this is another reason why physics must be careful not to gamble away its intellectual authority while playing with fantasies more suited to the science fiction industry. There are lots of things to do to understand our world theoretically, and to conserve our world practically.

I hope that the fascination for the universe and its many intriguing puzzles continuously accompanies you on the pages that follow. Doing scientific research is a privilege of the era we are living in. It is a big responsibility. That is why I wrote this book.

While completing my German manuscript, I discovered Sheilla Jones's *The Quantum Ten,* an excellent history of quantum mechanics. What really impressed me was how persuasively she explained the way modern theoretical physics has gone astray since the late 1920s. I am happy that we could write this book with such a common view, and I am grateful that her investigative mind as a journalist and her profound knowledge about history contributed to this restructured English edition.

Part I

SHORTCUT

Chapter 1

NOT TOO BAD, *HOMO SAPIENS*, BUT ...

REASONS FOR DOUBT: SOMETHING IS ROTTEN IN THE STATE OF PHYSICS

Enthusiastic applause rang through a crowded conference room in a Virginia hotel. Everybody gazed at a screen, where nothing but a simple diagram with a curve going through a couple of points could be seen. Only strange people could get carried away with emotion from something like this—like physicists at the annual meeting of the Astronomical Society, who continued to clap for several minutes.

What had happened? The plotted data confirmed with unprecedented accuracy a fundamental law of Nature: the emission of radiation from hot bodies. Discovered in 1900 by Max Planck, it was now lighting up the astronomy community with mathematical clarity. Even more spectacular was the origin of the data: microwave signals of different frequencies that did not come from a laboratory on Earth, but from a hot primordial state of the universe! A fireball of hydrogen and helium—without the molecular structure that would, in the distant future, make life possible—had released its light. More than 10 billion years later, it was picked up by the detectors of the man-made COBE satellite that had transferred the data just a few days earlier.

Replaying this story gives me the chills, as if I can actually feel the extremely cold temperature of cosmic radiation. It has a uniform distribution in space, which tells us that we should not delude ourselves that we live in a special place in the universe. Intelligent aliens could have come into existence everywhere! If they happen to look over our shoulders from time to time—an unlikely case—they would certainly have nodded their big heads appreciatively that afternoon of January 13, 1990.

SEEING THE LIGHT, OR ENLIGHTENMENT YET?

But *Homo sapiens* is not the most humble of creatures. During a NASA press conference, George Smoot, the project leader of the COBE satellite, called a picture of the cosmic microwave background radiation "the face of God," taking quite a faith-based tone. "Oh man, come back down to Earth," might have thought John Mather, his more modest co-laureate of the Nobel Prize. Surprisingly, however, such exaggerated language has spread in recent years, in particular among theoretical physicists. "The mystery of creation," as the famous cosmologist Alan Guth puts it, "is not such an unsolvable riddle any more. We now know what happened 10^{-35} seconds after the Big Bang." He appears to know it precisely.

Not only out in the cosmos, but also in the microcosmos of elementary particles, physicists are feverishly excited over such prospects. "No one could have imagined in his wildest dreams that we would get where we are now," says the theorist Brian Greene, in all seriousness. "Astronomers Are Deciphering the Book of Creation," "Physicists Close in on the Theory of Everything," and similar headlines appear in the newspapers. Such optimism appears even in distinguished scientific journals. But have our technological achievements really come at the same time that we have begun to understand the whole universe? That would be a strange coincidence. Do the theories that we are so vocal about really reflect what our eyes perceive?

It seems that today, almost 15 billion years after the Big Bang, we are about to solve the key questions about our universe. But this is a book of doubts...doubts as to whether the current theories in physics are actually close to the ultimate truth. We believe we know almost everything, but only one thing is certain: we currently live in an age that provides the best opportunity ever to look at the universe in marvelous detail.

The cosmic microwave background data is only one part of a revolution that has taken place in astronomy in the last few decades. Satellite telescopes are the cataract surgery that took away the murky and flickering atmosphere that constrained what astronomers could see. But a still more dramatic improvement came from digital image processing. This was a revolutionary innovation, like the discovery of photography, or even the invention of the telescope itself. How would Galileo, Kepler, Newton, and Einstein have enjoyed the present day! They surely wouldn't simply look up the most fashionable contemporary theories. Rather, they would look up to the skies to test their own. We have taken a giant leap out of our own solar system, out of our own galaxy, and deep into the universe through the use of precision telescopes.

THE NEW AND THE OLD

King Friedrich Wilhelm IV of Prussia once teased his royal astronomer: "Well, Argelander, anything new happening in the sky?"Argelander responded, "Does Your Majesty already know the old things?"

While you read this book, I would like to invite you to enjoy a few old tricks of observation that will help you appreciate the spectacular results of astrophysics and understand its riddles. The color of the light that hits your retina, for example, tells you something about the motion of stars and galaxies. The color of an object in space that is shifted toward blue light—with its higher frequency—indicates an object moving toward you, just as the sound of an approaching ambulance's siren appears to have a higher tone. If the original color is shifted toward the red end of the spectrum instead, it can be concluded that there is a receding motion. This phenomenon is called the "Doppler shift," and it applies similarly to both sound and light waves.

The Doppler shift in moving galaxies reveals that only a small fraction of the matter in the universe is visible through our telescopes. This was first noted as early as 1933 by Fritz Zwicky, a pioneer in galaxy research. He had a particular fondness for the Coma cluster, an impressive conglomeration of galaxies 300 million light years away. He measured the color of the objects in the cluster, and accordingly wondered about their high speeds. Their velocity, as shown by the Doppler shift in their emitted light, should have allowed them to overcome the gravitational attraction of the mass in the cluster, much like rockets escaping the gravity of the Earth. Therefore, in such clusters, additional mass had to be hidden, which, though invisible, prevented the galaxies from escaping the iron hand of gravity. Nowadays, we call this important discovery "dark matter," and it is a cornerstone of our current view of the cosmos.

Zwicky deserved the Nobel Prize for his discovery, but he tended to be unpopular with his colleagues because he was blunt and bullheaded. However, he really shot himself in the foot by arguing against *the* cosmological discovery of the 1930s, made by his great rival, Edwin Hubble. While measuring the redshift of the light of the galaxies, Hubble observed that almost all of them appear to be moving away from us. The more distant they were, the faster they receded.

Fig. 1. Left: Coma galaxy cluster 300 million light years away. Right: Fritz Zwicky, a pioneer of galaxy research.

Redshifted galaxies were the first piece of evidence that showed we live in an expanding universe, which we now attribute to an explosion-like process in the early universe…the Big Bang.

WHY THE EYES OF ASTRONOMERS TWINKLE

The applause in that crowded conference room in Virginia in January 1990 was a celebration of the Big Bang model, too. The cosmic background radiation data, which echoed the early phase of the universe, confirmed that the universe was hot and dense back then. This favored the idea of a continuous expansion since the Big Bang, which was born with Hubble's observation of redshifts. Later, the formation age of the cosmic microwave background radiation was determined to be 380,000 years after the Big Bang. In comparison with the next 14 billion years, that's quite a short period of time. Science has never been that close to the eye of the storm! Thus, the Big Bang is now generally believed to be the beginning of the universe.

There would be more applause to come. In 1998, two research groups caused further excitement about the expansion of the universe after analyzing images of supernovae, very bright star explosions. Their findings were worthy of the 2011 Nobel Prize in Physics. In short, the modern version of Edwin Hubble's measurements showed that the universe is not just expanding, but that the expansion is accelerating, thereby contradicting all reasonable expectations. The whole paradigm of cosmology was turned upside down, and the data seemed to require the existence of an entirely new concept—a "dark energy." This is essentially a force that has repulsive gravity, pushing all the masses in the universe farther away from each other and at an ever-increasing pace.

Isaac Newton famously said in 1687, "Gravity is the natural phenomenon by which physical bodies appear to attract each other." This now appears to be superseded by the discovery of dark energy. Does the new concept now deliver *the* complete picture of the cosmos? The precise data has undoubtedly led to new insights, but it has raised more questions than answers.

THE ALLURE OF GEOMETRY

While contemplating the parabolic shape of Sugarloaf Mountain at the Praia Vermelha in Rio de Janeiro, I noticed street hawkers offering fresh coconut milk right from the nut. Artfully, they made a triangular cut on top of the nuts. The slightly curved edges opened the way to the inside of the fruit, but due to the round shape of the coconut, they were almost at right angles to each other. I didn't hesitate to get one, as it was the perfect prop for my upcoming presentation. At big physics conferences, sometimes you have to resort to unusual things in order to get attention. I was attending the Marcel Grossmann meeting, named after a Hungarian mathematician and friend of Einstein, which took place in Rio in 2003. Gravitational physicists and astronomers had come together from all over the world to exchange their ideas.

Unfortunately, my talk was scheduled for the late afternoon. The physical and mental presence of the audience declines greatly at this time of day. So I was glad to have the coconut and the straw, which I used to illustrate an important concept of the theory of general relativity. The angles of a flat triangle must always add up to 180 degrees, but triangles on a curved surface can easily have angles with a sum exceeding 180 degrees. The coconut hawker had cut three right angles, making a triangle with 270 degrees. No matter how you may shift the straw along the edges (mathematicians used to analyze curved surfaces by doing this), the angles won't change.

Such problems sometimes incite heated discussions, maybe because general relativity, with its geometrical abstractions, touches the deep emotions of physicists. The idea that everyday objects can demonstrate space-time curvature is utterly fascinating. The Russian Nobel laureate Lev Landau, the author of a brilliant ten-volume series on theoretical physics, commented on this fact. He wrote that Einstein's great achievement of 1915, the subtle geometric refinement of Newton's law of gravitation, is "the most beautiful physical theory." At the time I was in Rio, I was blissfully unaware that general relativity would be called into question by new observations.

I was part of the majority of physicists who rather suspected the skeptics to be ignorant of the second volume of Landau's book series, where the theory was explained in a short but concise way.

FROM THEORY TO OBSERVATION

Fortunately, not only theoreticians but also many observational astronomers had descended upon Rio. During my flight from Madrid, I sat next to a PhD student from Naples named Sante Carloni, and for the entirety of the nine-hour flight we geeked out over physics. Maybe it was the flight above the clouds that led us to the subject of slight deviations of the trajectories of the Pioneer spacecraft, on which a detailed study had been published in 2001 by NASA.[1] Sante told me that the researchers had determined a slightly larger value for the spacecraft's acceleration toward the Sun than was to be expected according to Newton's law. This subsequently drew a lot of attention to low-acceleration tests of gravity.

During the conference, Sante and I became fast friends. He still held an untenured position, and didn't look down his nose upon seeing the badge I was sporting, which showed a Munich high school as my institutional address. Rather, it reminded him of his own schoolboy pranks. The Rio conference took place at a military academy, and straight-laced, tough-looking guards were ensuring that everybody showed the appropriate discipline. Next to the main entrance of the academy, Sante and I were discussing a coincidence between the Pioneer acceleration and dark matter at the edges of galaxies, when we were rudely interrupted by a uniformed guard. We were not supposed to be sitting on the wall surrounding the flower beds. A suitable quote by Einstein about those who joyfully march to music in rank and file came to our minds. However, we considered it more prudent to continue our discussion elsewhere.

If dark matter was indeed related to the low accelerations felt by stars in the outer parts of galaxies, something had to be wrong with the whole concept of dark matter. We agreed on that, and Sante suggested we attend a session that would be presenting the most recent measurements. One speaker reported the percentages of the three components of the universe—dark energy, dark matter, and ordinary matter—to be 72, 25, and 3, respectively. I already felt some unease with the concept of dark matter and dark energy, and the claimed accuracy seemed exaggerated to me. Since my freshman years in college, astronomers with their huge inaccuracies were considered the grubby kids of physics. Due to their cryptic corrections and missing error budgets, they could only dream of reaching the precision shown in fields like quantum optics.

I teased Sante, "You astronomers of all people! Some ten years ago, you didn't know whether the universe is 10 or 20 billion years old, and now you are measuring accurately to one percent!" He defended himself with some southern Italian swear words about theorists, but then agreed frankly. "Well, dark energy," he said, "or call it quintessence, that's just naming something we haven't the faintest idea about." During my flight back to Germany, thoughts on dark substances continued to tickle my brain.

A COSMOLOGY OF COLLECTING DATA

Shortly after returning to Munich I found myself at a very different meeting—one of the regular assemblies of teachers at my high school. Unfortunately, the inevitable bureaucratic issues sapped my attention span, but I kept wondering about the nature of dark matter and dark energy. Quite a lot of things seemed questionable, if not contradictory. Even the students in my astronomy course often remained dissatisfied with the superficial explanations. They were asking the right questions. What could those dark substances consist of? Why the hell did Nature invent them in the first place?

Many astronomers, in contrast, seem to believe it's their purpose simply to measure the respective percentages precisely, keeping to the motto: "We don't know anything, but at least we know it accurately." There are few thoughtful voices like that of Anthony Aguirre from UC Santa Cruz: "Although this paradigm has received considerable support from recent observations, this support has been at the expense of simplicity..."[2]

The current paradigm, the "concordance model" of the universe, is supported by most cosmologists. It is described by six numbers in total (two of which are dark matter and dark energy), and you'll become more familiar with them as the story progresses. But this is the question that really bothers me: Why is the universe as a whole described by these six numbers, and not by others? And why just six? Are we able to accurately calculate them, and if not, why not? And what happens if new, still more precise observations come up with even more unexplained numbers?

BIG EXTRAPOLATIONS AND LITTLE DOUBTS

If you are out to describe the truth, leave elegance to the tailor.

—*Nikola Tesla, American physicist*

The more I read up on gas clouds in galaxy edges and on X-ray radiation from clusters of galaxies or distant supernova explosions, the more

I wondered whether one has to interpret these observations by positing dark substances. Should we cast doubt on the accepted theories of physics instead? Combing through the library, I found a well-known textbook on galactic dynamics where the authors state:

> It is worth remembering that all of the discussion so far has been based on the premise that Newtonian gravity and general relativity are correct on large scales. In fact, there is little or no direct evidence that conventional theories of gravity are correct on scales much larger than a light year or so. Newtonian gravity works extremely well on scales of 10^{12} meters, the solar system. (...) It is principally the elegance of general relativity and its success in solar system tests that lead us to the bold extrapolation to scales 10^{19}–10^{24} meters...[3]

Wow! Fancy that. Two leading experts claim that the law of gravity has been well tested in our solar system only—a tiny fraction of the universe that corresponds to a single snowflake in all of Greenland. Scientists seem drawn to the "elegance" of the theory, which is not really a scientific criterion. I often confront physicists and astronomers with this quote. Usually they shrug and reply airily, "That is indeed true, but why shouldn't the law of gravity be valid? So far, there is nothing better to replace it."

I do not claim that the intelligent aliens—maybe they don't exist and we shouldn't worry—have better theories, but I do believe one thing. Our confidence in having found a definitive "concordance model" of cosmology would make them smirk at our naïveté. And I am also pretty sure that this smirk is an innocent mien compared to the sardonic grin that might appear on their faces upon looking at other theories of *Homo sapiens*.

COSMOLOGY IS RELATIVELY OK

Despite all drawbacks, cosmology is still in good shape when compared to the sum of findings physicists have gathered from observations of the microscopic world—the standard model of particle physics. Cosmology's "concordance model" uses six numbers, which are called "free parameters" because they cannot be explained within the model but rather are fitted to the measurements. The standard model of particle physics needs not only six of them, but an impressive 17. Why 17? The above questions about the six numbers in cosmology are dramatically amplified when it comes to particle physics.

The story of these numbers began with great expectations similar to those we have for today's telescopes. In the 1950s, a boom of particle accelerators started producing hundreds of elementary particles with spectacular collisions. In the following decades, particle physics has been busy classifying this zoo and reducing its mathematical description to "only" 17 parameters. A few Nobel prizes have even been handed out because of this work. But should we be convinced that we have come to understand the ultimate structure of matter? In his book *The Trouble with Physics*, Lee Smolin comments on the 17 free parameters. "The fact that there are that many freely specifiable constants in what is supposed to be a fundamental theory is a tremendous embarrassment."[4]

If we naively add up 17 and 6, we would need 23 arbitrary numbers in all to understand physics and cosmology together. But it seems that physicists can't keep track of the count anymore. The British cosmologist Martin Rees and the Nobel laureate Frank Wilczek, in a review article,[5] counted a total of 31 fundamental constants, all of which are, again, unexplained numbers. Interestingly enough, at the same time, Rees published a popular book with the catchy title *Just Six Numbers*, referring to the parameters in cosmology. As you can see, selling books on cosmology is still possible, but imagine trying to offer a treatise on physics to the public with the title *Just Thirty-One Numbers*!

> A theoretical construction is unlikely to be true, unless it is logically very simple.
>
> —*Albert Einstein*

THE ODD COUPLE

Such a bunch of free parameters that purports to accurately describe Nature reminds me of a quote by King Alfonso X of Castile. "If the Lord Almighty had consulted me before embarking on creation thus, I should have recommended something simpler." He made that comment when he came to know of the Ptolemaic epicycles, an incredibly complicated description of the planets' motion based on an Earth-centered view. This theory was later thrown into the junk heap of science by the Sun-centered model of Newton's law of gravity.

Physics is the science of explaining the world simply and logically. One single gravitational constant determines the orbits of thousands of celestial bodies in our solar system. Hundreds of spectral lines, the fingerprints of atoms that explain the whole system of chemical elements, are described by quantum mechanics as electron waves. This is determined by using a sophisticated technique, which boils down to the same math one uses to measure how springs—just like your Slinky—oscillate in three dimensions. In contrast, today's particle physics and astrophysics seem to be much more complex. But does this mean that further changes that may greatly simplify our view of the world are impossible? Where did scientists gain the confidence that our two-pound brains have already found the ultimate theoretical framework?

Everybody agrees that there is a grave problem in fundamental physics. Though Einstein's general relativity theory is as successful in describing the universe on large scales as quantum mechanics is in explaining the properties of atoms, the respective mathematical formalisms are desperately incompatible.

For example, the gravitational attraction between a proton and an electron is smaller by a factor of 10^{40} than the electrical attraction that binds these two particles into a hydrogen atom. That's a number with 40 zeros in it, which is no small difference. But physicists have yet to explain why this large difference between the two forces exists. That's a lot of zeros to ignore! In a textbook on quantum field theory, I found a comment that amazed me: "In view of this, the particle physicist is justified in ignoring gravity, and because of the huge theoretical difficulties, he is happy to."[6] Sometimes, physicists behave like drunks in search of a lost house key. They search around the base of a street lamp, even though they know that they lost the key somewhere else in the dark.

If the law of gravity really needs to be corrected, a cascade of corrections in other fundamental aspects of physics would be necessary. The cognitive dissonance with regard to such basic problems is an irritating phenomenon in contemporary science. It's there to see when you study old physics publications. The mysterious factor 10^{40} puzzled the famous British physicists Arthur Eddington and Paul Dirac some 75 years ago. Such gems are swamped by the mass of today's academic papers, but it is really worthwhile to dig them up. Original articles by Einstein, Dirac, and Feynman are classic hits, and their quality is not based on their Billboard rating.

PLAYING TECHNO WITH STRING INSTRUMENTS

Now, let's check out something hip. For the last three decades, a possible solution to the conflict between quantum mechanics and general relativity has been advocated by "string" or "superstring" theory. It claims that particles are not pointlike, but oscillations of tiny strings that are supposed to pervade space-time. Also, in addition to the three visible dimensions of space, string theory assumes other dimensions rolled into structures that are too small to be directly observable. Although there is no experimental evidence yet to be had, many physicists are more than happy to dance to the tune of string theory and its promises of unifying all theories. The physicist Brian Greene states in his book *The Elegant Universe*,

> According to superstring theory, the marriage between the laws of large and small is not only happy but inevitable.... String theory has the potential to show that all of the wondrous happenings in the universe...are reflections of one grand physical principle, one master equation."[7]

All this would be surely more impressive if string theorists had an idea about how this equation looks. In Greene's follow-up book, *The Fabric of the Cosmos*, there is still more enthusiasm that, sadly, didn't jump over to me.

Or maybe, in spite of having read such books, I still do not understand very much of string theory. If you share this experience, you are not that alone. Günther Hasinger, the director of the Institute for Astronomy at the University of Hawaii and recipient of the US$3 million Leibniz Prize, also hasn't the faintest clue when it comes to understanding string theory. At least, that's what he says.[8] Now, Hasinger isn't a household name, but even the most brilliant postwar physicist, the Nobel laureate Richard Feynman, had this to say about string theory:

> I know that other old men have been very foolish in saying things like this, and, therefore, I would be very foolish to say this is nonsense. I am going to be very foolish, because I do strongly feel that this is nonsense!...I don't like that they're not calculating anything. I don't like that they don't check their ideas. I don't like that for anything that disagrees with an experiment, they cook up an explanation—a fix-up to say, "Well, it still might be true."

WHERE THERE IS MISERY, HOPE FLOURISHES

Where do these extreme opinions come from? As a matter of fact, there is not yet a single piece of experimental evidence supporting the assumptions of string theory, nor a proposal by theorists as to what exactly needs to be measured that might provide supporting evidence. Instead, the theory—some say that it is only a hodgepodge of ideas resembling a theory—is continuously admired for its marvelous beauty. String theorists rebut attacks that there is as yet no backing evidence at all for the theory by claiming that it is simply too elegant to be gotten rid of. The theoretical physicist and string theory "dropout" Lee Smolin listed in his book *The Trouble with Physics* all of string theory's shortcomings: its technical flaws, circular arguments, arbitrary concepts, and even the sociological component of what should be objective theory, such as the strange hierarchical structures in this "community" of physicists.

Smolin attacked string theory very methodically and respectfully. He never started playing hardball, though. His description of string theory's numerous faults would make you flush with anger. Too much of the story is reminiscent of the tale of the emperor's new clothes. Initially, I felt uneasy about being unable to fully appreciating the ever-growing mathematical complexity of string theory's high-dimensional Promised Land. Smolin helped me put a finger on just what I was justifiably uneasy about.

ASTROPHYSICS BECOMES INFECTED BY THE SPECULATION VIRUS

But what to do without string theory? If you talk to die-hard particle physicists, they'll talk your ear off, telling you how wonderfully the standard model stands up to experimentation. It particularly hurts when this story is drummed into the heads of the young, because as you now know, the success of the standard model is dearly bought by the necessity of incorporating 17 arbitrary numbers that the theory cannot explain. There are many others who simply can't stomach this selling-out of physics.

For a long time, there has been a brain drain from particle physics to cosmology, something that is easily accomplished if one repaints the office door to establish a group for "Theoretical Astrophysics."* Not coincidentally, one of the most beautiful complications of cosmology—a ubiquitous field in space called "quintessence"—we owe to a former particle physicist. And the director of the Center for Particle Cosmology at the University of Pennsylvania is working "firmly on the particle physics-cosmology border." The question remains whether this border exists.

But isn't attracting these experienced researchers a big win for astrophysics? No. Their problem is that for decades particle physicists have become comfortable describing nature with arbitrary numbers instead of relying on basic principles, as Einstein always did. Those who have forgotten to wonder are easily satisfied by superficial concepts such as dark matter and dark energy. Still worse, on the Internet almost every week you can see similar "fields" conjured up in order to "explain" some new contradictory cosmological observation. The standard reasoning: "We know that in micro-physics that they exist, too."

At the risk of annoying more than a few people, I would say that our understanding of the universe has not advanced due to this guest performance by

---

* For example, at Fermilab, the biggest US accelerator.

particle accountants. Rather, particle physics has suffered from a bubble of meaningless numbers, and this inflation is bleeding over into cosmology, dooming it to the same fate. Particle physicists have exported their crisis to astrophysics by trying to sell it their worthless "free parameter" currency. There is, sadly, no sign of a bailout in the near future.

There are still string theorists who have not yet replaced their empiricism with ideology, but their theories are not one iota better. To prevent their ideas from being disproved, they postulate phenomena at the so-called Planck scale of 10^{-35} meters. Keeping in mind that the size of the smallest stable structure (the atomic nucleus) is 10^{-15} meters, it is silly to hope for an experimental confirmation of the Planck length. The Planck length is smaller than the atomic nucleus by the same powers of ten as the atomic nucleus is smaller than Earth.

Even more disconcerting is that if something is wrong with the law of gravity or the gravitational constant G—and there is evidence suggesting this—then the entire construct of the Planck scale based on G would turn out to be complete baloney.

PHYSICS BECOMES SCIENCE FICTION

A good novel tells us the truth about its hero; but a bad novel tells us the truth about its author.

—*G. K. Chesterton, English writer*

A concept called "inflation" has been tacked onto the standard model of cosmology. Inflation allegedly started when the universe was about the size of that very tiny and unmeasureable Planck length. This offers a cushy niche for theories that would not survive in the harsh climate of experimental measurements. And high-energy physicists are increasingly aware that the Large Hadron Collider at CERN in Geneva is likely to be their last big taxpayer-funded toy. They then plan to seek refuge at a less expensive accelerator: the Big Bang, a unique laboratory for phenomena that are unobservable on Earth. It's just too bad that you can't observe what happens at the Big Bang either. However, this serves only to inflate the imagination. For instance, a *Scientific American* article about cosmic inflation and string theory states that the universe may be "on a membrane, sitting at the tip of a spike or composing part of a membrane wrapped around tea-cup-like handles" or may look like "an octopus with several elongations."[9] Accordingly, the cover story is illustrated by an image of something that looks like a cross between a roller coaster and an HIV virus. Scary, but the authors conclude their ludicrous tale with

words of comfort: "Fortunately, our universe does not have to be annihilated to benefit from this inflationary process."

I know many physicists who are annoyed by such malarkey, but most
of them are working in laboratories or in the field of applied physics.
"Fundamental" physics, however, cultivates this style of fancy speculation, without any data, any observation, and any experiment. It postulates a litany of increasingly bizarre concepts: hypothetical new particles,
wormholes, cosmic strings, membranes, primordial black holes, eternal or
chaotic inflation, warped space-time, supersymmetry, extra dimensions,
quantum foam, and multiverses.

HOMO SAPIENS, IS ANYONE HOME IN THERE?

Recalling our vigilant aliens, we might suppose that at this stage they would
send us a therapist. Besides being concerned about our highly exaggerated
fantasies and the considerable disconnection from reality, our alien friend
would probably diagnose us thusly: we suffer from the hallucination of having understood the ultimate laws of Nature, combined with the urgency to
communicate it. According to a Chinese proverb, one fool can affirm more
nonsense in one day than seven wise men can disprove in a year. This is
a methodological problem for many sciences. But there was a time when
physics was different. Where has Niels Bohr gone, whose intuition guided
the physicists of the 1920s to the discovery of quantum mechanics? Where
are such geniuses as Einstein, Heisenberg, Dirac, and Feynman today?

When comparing the advances made by such historic characters, even
the luminaries of the recent past, such as Steven Weinberg or Stephen
Hawking, glow only faintly, not to mention those mayflies whose sirens
(with a backup string orchestra) we are listening to today. They fiddle with
the most complex calculations, and don't realize that they lost their scientific grip long ago.

Compared to earlier periods, today's lightweight careerists lack not only
Einstein's brain, but also his backbone. They are merely interested in gathering funding, networking, and scientific lobbying, and seem to have lost
interest in figuring out the innermost workings of the universe. Instead of
climbing a rock to contemplate, as Heisenberg did when he came up with
his key contribution to quantum mechanics in 1925, scientists are happy
with their mutual applause at conferences. Physicists have cheekily rede-
fined arrogance as avant-garde. And as a last resort, computer models fed
with teraflops and terabytes flimsily bolster their virtual physics, making
such theories seem untouchable by old-fashioned experimentation, the
hallmark of good science.

The taxing load of doing day-to-day physics is usually carried by scientific sherpas, working on temporary contracts, who are given little chance to reflect on what is sold as ultimate truth by the trendsetters at Princeton and Harvard. Who wants to rub those guys the wrong way?

Today's physics has lost its idealist thinkers, and its huge institutions are drifting tankers lost at sea with nothing aboard but their vanity, funding necessities, and the market rules of science. A generation of physicists has lost any sight of land. The geniuses are dead, and the successors are devoting themselves to excessive theoretical speculations while overselling their nonresults. Physics has entered a phase of degeneration.

> Very strange people, physicists—in my experience the ones who aren't dead are in some way very ill.
>
> —*Douglas Adams,* The Long Dark Tea-Time of the Soul

Chapter 2

GALILEO WOULD FREAK OUT!

A QUANTUM LEAP IN MEASURING DEVICES: WHY WE LIVE IN FANTASTIC TIMES

There are many reasons to doubt modern theorizing in physics, but this modern era—400 years after Galileo first peered into the universe—has become an era of extraordinary observations. New techniques allow us to "see" the universe in an unprecedented way. We've come a long way since Galileo first directed his simple little telescope at the skies.

In the fall of 1610, Galileo said obliquely, "The mother of love imitates the shape of Cynthia." By Cynthia he meant the Moon, whose changing shape through the phases from new moon to full moon has been known to astronomers for several thousands of years. Galileo then noticed a similar periodic blackout of the "mother of love," the planet Venus. This was the vital blow to an already-frail geocentric belief system.

Galileo's observations were made possible by the work of Dutch inventors who built looking glasses that magnified normal vision by three times. Their patent request described the instrument as a device for "seeing far-away things as though nearby." But Galileo, working in Italy, understood the construction still better and quickly optimized it to an apparatus that deserved the name "telescope." Progress in science is often the result of the meshing of physics and technology. This particular leap made by Galileo was possible because researchers had already developed an understanding of light refraction through glass.

Soon afterward—astronomically speaking—the Hubble Space Telescope was orbiting Earth at an altitude of 350 miles, and it was taking pictures at a resolution 500 times better than Galileo's telescopes and 1,000 times more sensitive than the human eye. Of course, the Hubble Space Telescope would not have existed without the groundwork prepared by nineteenth-century pioneers of electrodynamics such as Faraday, Ampère, and Maxwell. And without the discovery of electromagnetic waves by Hertz, we would never have been able to see the telescope's unique photographs sent back to us by radio transmission.

THE UNIQUE CO-EVOLUTION OF PHYSICS AND TECHNOLOGY

There is a nice historical interplay between theory and practice in the development of satellites. Galileo's observations were complemented by the Copernican Revolution—the shift from a geocentric to heliocentric model of the universe—all the way up to Newton's theory of gravity. Without an understanding of the laws of planetary motion and gravity, advanced rocket and satellite technology would simply never have developed. Nowadays, astronomers use space telescopes to test Newton's laws, achieving an unprecedented precision.

Along with gravitation and electrodynamics, there is yet another pillar of physics that plays an important role in our ability to "see" the universe—quantum mechanics. We benefit from Einstein's description of the photoelectric effect back in 1905: he understood that a quantum of light (a photon) could kick out an electron from a metal plate. Much later, this insight translated into the technology of a digital camera with millions of detectors for photons,* effectively being a high-tech replacement for the human eye. Astronomers harvested the fruit, and computer graphics allowed another tremendous progress in analyzing the data of the starry sky.

QUICK SHIFTS OF ASTRONOMY

One of the first people to couple a telescope and a CCD detector was Rudy Schild from the Harvard-Smithsonian Center for Astrophysics. I got to know him at a cosmology conference that took place in St. Petersburg in

---

* American scientists Willard Boyle and George Smith from Bell Labs won a Nobel Prize in 2009 for inventing the technology of charged coupled devices (CCDs) in 1969.

Russia during the "white nights," the height of summer, when it doesn't quite get dark at night. Rudy gave his lecture on his favorite topic—quasars, which are dot-like bright objects, almost like stars, but far beyond our galaxy.

Although Rudy's first CCD camera produced very low-resolution pictures at 100 x 100 pixels, they were nonetheless helpful for analyzing these exotic objects. At the time of their discovery, the quasar's light composition was a great riddle, and astronomers were shocked about how very far away they seemed. Astronomers today assume that quasars are cores of young galaxies with luminosities many times higher than those of their grown-up kin. It is the fluctuating brightness of quasars that scientists find especially fascinating. Galileo got rid of the notion that the sky has always looked the same, and since then, astronomers get excited about objects that change.

Rudy recounted how one night he met a group of quasar observers talking excitedly.

"Look, it's lighting up!"

"Yes indeed, and now it's getting darker again. Wow!"

Finally someone asked him, "Hanging around? What are you doing here?"

"Well," said Rudy, "I was observing, too. But as high clouds showed up, I thought that I should close the telescope dome."

After a moment of silence—overhead clouds, too, cause a variation of brightness—everybody laughed and went to work again the following night, full of enthusiasm. This anecdote highlights how much the astronomer's work is eased by telescopes in satellites, which orbit high above any cloud cover and enable the undiluted observation of the highly fascinating glowing of quasars.

Astronomers can estimate the size of a quasar by the brightness of its oscillations. Since no mechanism within the quasar can spread faster than light, a fluctuation once every 24 hours indicates that the size might be one light day, the distance light can travel in the meantime. This is extremely small, considering that galaxies with similar luminosities usually have a diameter of 100,000 light years. Hence, quasars show a dense concentration of matter, which initially led astronomers to believe that these objects were much smaller and closer than they turned out to be. All this was discovered a few decades ago, and nowadays automatized telescopes have identified tens of thousands of quasars, providing important information about the early universe.

SLOW SHIFTS IN THE STARLIT SKY

Galileo's observations of the phases of Venus and the constellations of the Jupiter moons were revolutionary because they contradicted the dogma about the constancy of the celestial sphere. In a subtle way, we have a similar problem today. We know very little about possible changes in the cosmos. All the information we have on the dynamics of outer space is what we have to conclude from a "snapshot" that is only a few decades long.

If one relates the age of the universe—14 billion years—to one year of observation, a change in the laws of Nature will show up in the tenth decimal place at best—barely discernible. In order to look over Nature's shoulder and detect such tiny changes in the dynamics of the universe, we need extremely accurate clocks.

There are natural cosmic clocks. Pulsars were discovered in 1967 due to their strikingly fast and regular pulses of radio signals. For a while, researchers joked about "little green men" sending signals to Earth. Then it became clear that in reality the pulses were generated by neutron stars, the rapidly rotating remnants of supernova explosions of stars that had collapsed into spheres of about ten miles across.

Since then, atomic clocks have caught up with the accuracy of pulsars, and physicists are about to build clocks that are accurate to the seventeenth decimal place. The more precise the measurement, the easier we can detect surprising deviations in the laws of Nature.

MEASURING THE WORLD ANEW

As important as it is that we measure time, we must also measure space accurately if we want to understand the world around us. Lasers are great tools for precise distance measurements, and this is just one among many examples of how our knowledge benefits from sole technological progress. For instance, in the Gravity Recovery and Climate Experiment (GRACE) mission, two satellites—nicknamed "Tom" and "Jerry"—are orbiting Earth and recording their distance accurately to microns. The satellites fly at the exact same altitude, so the distance from each other can only be affected by irregularities in the Earth's gravity. Evaluating the data yields an impressive map of the local gravity and its tiny variations.

A similar idea guided an even more ambitious project, the Laser Interferometer Space Antenna or LISA, funded by the European Space Agency (ESA) and NASA. In this case, three spacecraft were to be positioned in an equilateral triangle in the Earth's orbit around the Sun, with each arm measuring 3 million miles. The plan was to detect gravitational waves (e.g. radiated by pulsars) by measuring that gigantic distance to the accuracy of a fingernail. Moreover, if there was the slightest deviation from Einstein's general theory of relativity, LISA would be able to detect it. The launch was scheduled for 2022, but NASA withdrew from the project in 2011. ESA is now preparing a scaled-down version.

Taking measurements from satellites in space means that scientists can escape from the various motions of the Earth's surface. The Earth's crust can be significantly distorted by tidal forces, changes in atmospheric pressure, and even ocean waves. Such movements affect the accuracy of the distance measurements to the Moon that use lasers. In 1969, Apollo 11 astronauts Neil Armstrong and Buzz Aldrin deployed an array of mirrors on the Moon, with more added in later missions. The mirrors are targeted by powerful lasers, and then the time it takes for light to go from the Earth to the Moon and back is measured, allowing the distance to be determined accurately to less than an inch. This pretty impressive technique is called "Lunar Laser Ranging." There are a number of research stations around the globe doing this, and researchers are using the distance measurements to test such things as the constancy of the gravitational constant G and various predictions of Einstein's theory of general relativity.

THE SKY ABOVE US AND THE GROUND BENEATH US

I regularly bring my astronomy students to the Laser Ranging station in the Bavarian Forest in Germany. I think hands-on experience is essential if you want to get young people fascinated about science. The observatory is located in the town of Wettzell, where you can also go on amazing hikes in the forest.

The Moon is a moving target because it travels almost one mile in the time it takes the laser beam to travel from the Earth. When researchers conduct Laser Ranging at the station, they are happy every time they "hit," even if all they get reflected back is just a couple of photons from the moon or from satellites designed for the sole purpose to test the law of gravitation, the 900-pound mirror balls called LAGEOS (Laser GEOdynamic Satellite).

The Laser Ranging station at Wettzell has an impressive collection of precision instruments, such as a ring laser that measures the Earth's rotation and a gravimeter that can determine the local gravity accurately to the tenth decimal place. You do, however, have to keep a distance of several meters from the instrument so as not to disturb it. One of our visits still manifests itself in their raw data!

Fig. 2. The Radio telescope in Wettzell was just being freed from the snow in the cozy February weather. The name "fundamental measuring station" comes from the fact that it measures the basis for the space-time coordinate system of the Earth.

The device at Wettzell is part of a network of radio telescopes spanning the world. These devices receive low-frequency electromagnetic signals with wavelengths of up to several meters. The angular precision with which these signals are received has improved greatly over the last decade. Why is that? You first have to understand what happens if you take an amateur astronomer's telescope and cover half the lens. Will the visible field be cut in half? No. Our astronomer will continue to see everything. The light intensity will have decreased, but the angular accuracy will remain almost the same.

One can reverse this effect and increase the precision with which you can spot an object by interconnecting several separate telescopes. You don't collect much more light, but the large ground distance of the interconnected telescopes greatly improves the angular resolution. Using a trick called Very Long Baseline Interferometry (VLBI), radio telescopes can effectively span the size of Earth! This system of telescopes is used for the precise determination of both quasar positions and the Earth's coordinate system. VLBI measures tiny changes in the Earth's rotation axis (e.g., such as those that occur during earthquakes) and millisecond variations of the length of the day. Again, this is where extremely precise clocks come into play for the evaluation of this data.

STUMBLING UPON THE MILKY WAY AND THE BIG BANG

For a long time, radio waves were considered to be man-made, and antennas were used for telecommunication only. That changed in 1931 when Karl Jansky, a physicist at Bell Labs in New Jersey, detected a small but peculiar signal glitch in a transatlantic radio link. The signal kept repeating, slightly less than 24 hours apart. An astronomer explained to Jansky that the Earth rotates in exactly 23 hours and 56 minutes with respect to the starry sky (not to the Sun!), and therefore the signal obviously originated outside our solar system, from the direction of the center of our galaxy, the Milky Way. Later it was discovered that many quasars, being galaxy nuclei, emit radio waves as well.

It just so happens that such an accidental discovery repeated itself 34 years later. That's when the Bell Labs physicists Arno Penzias and Robert Wilson were disturbed by a signal glitch while building a low-noise antenna. This unpleasant effect turned out to be the first evidence for the cosmic microwave background, commonly viewed as the echo of the Big Bang.

Yet again, somebody else, the astrophysicist Robert Dicke this time, had to explain to these lucky fellows what they had actually found. Dicke had been building an antenna to look for the microwave background radiation himself. However, he had been doing it on purpose! Still, it took another 25 years before the cosmological significance of cosmic background radiation was made apparent through the data from the NASA Cosmic Background Explorer (COBE) and the more recent Wilkinson Microwave Anisotropy Probe (WMAP) space missions.* Jansky was honored with a physical unit (radiation flux) being named after him, while Penzias and Wilson received the Nobel Prize. It's a little irony of history that both the astronomer who had pointed Jansky in the right direction and Dicke, who had already predicted the cosmic microwave background, were left empty-handed.

ACRONYMS "R" US: GREAT OBSERVATIONS WITH THE BIG OBSERVATORIES

One can easily get lost in the forest of acronyms when surveying the multitude of satellites and space probes that are delighting scientists with precise data of the universe. And they are providing data on a wide range of fronts.

From high-energy gamma rays all the way down to low-frequency radio waves, the electromagnetic spectrum contains a variety of "colors"

---

* Their successor, Planck, has just completed collecting even more precise data.

that are hidden from our limited range of vision, but it tells us astounding facts about stars and galaxies. Today, space telescopes cover almost every wave length,* thanks to the ambitious Great Observatories program initiated back in the 1970s by NASA. Today, the satellite telescopes Chandra, Fermi, and SWIFT, for instance, study the shortest wavelengths, gamma ray bursts, and X-ray emissions.

SWIFT practiced an interesting collaboration. Whenever it detected an intense burst in a certain region of the sky, the position was quickly emailed to thousands of amateur astronomers. The amateurs on the ground often managed to observe the famous afterglows of the gamma bursts in the visible light spectrum, making a valuable contribution to scientific research. This also highlights how science benefits from a connected world with open data.

There are ground-based detectors trying to unravel the enigma of ultra high-energy gamma rays, like the Pierre Auger Observatory in Argentina's western grasslands. It covers an area of the size of the state of Rhode Island, big enough to capture some of the extremely rare events that have an energy 1 million times higher than the CERN particle collisions. When the energetic rays smash into Earth's atmosphere, they trigger a shower of particle debris that is analyzed with arrays of little telescopes and delicate detectors on the ground.

Ground-based telescopes are typically located at high elevations to limit the blurring of images by atmospheric fluctuations—the reason we see stars twinkling. That problem has been significantly resolved by "lucky imaging," a technique that overlaps single pictures taken by a high-speed camera. "Lucky imaging" will also be used for the European Extremely Large Telescope (E-ELT) planned by the European Southern Observatory for Chile. It will have a main mirror that measures 40 meters across, which makes it the largest telescope in the world. By comparison, Edwin Hubble's legendary instrument in Mount Wilson Observatory in California was just 2.5 meters in diameter. The E-ELT, expected to be completed in 2020, will be able to observe a subtle change in the cosmological redshift, Hubble's ground-breaking discovery. He would certainly be happy about that! Another fantastic instrument that allows a look back into the early universe is the Acatama Large Millimeter Array (ALMA). Observing the sky at a distance from Earth remains essential, since the atmosphere masks a wide range of the infrared and ultraviolet spectrum. Without spacecraft missions, astronomy would be blind to these "colors."

---

*See "List of Space Telescopes," *Wikipedia, the Free Encyclopedia,* en.wikipedia.org/wiki/List_of_space_telescopes.

The next generation of space telescopes is now in the works. While the famous Hubble telescope whipped around the Earth every 97 minutes, the new James Webb Space Telescope (JWST), scheduled for 2018, will be sitting about 1 million miles out in space, as a little partner of Earth orbiting the Sun. Due to its new instrumentation, JWST will have the power to look deeper into space than ever before. We have, indeed, come a long way from the telescope of Galileo.

FROM HUNTERS AND COLLECTORS TO CULTIVATION

The rapid technological advances in telescope technology are not all that is changing for astronomers. The life of the researcher has been undergoing a sociological evolution. Astronomers such as Tycho Brahe (from whom Johannes Kepler got the data for his planetary laws), Galileo Galilei, Wilhelm Herschel, Edwin Hubble, and Walter Baade had one thing in common. They were all lone-wolf researchers who stuck close to their instruments on winter nights with clear skies as they meticulously recorded their observations. These astronomers obsessively pursued their big questions, but were also masters of their instruments. They had to correct the color of their raw images, care about Earth's rotation, and endure all other difficulties that astronomy offers.

In contrast, today's telescope technology is fascinating but barely manageable by a single person. There isn't the experience of staring directly into the night sky and having its splendor disclosed. Most observations at ESO's Very Large Telescope in Chile are run in service mode, whereby all mechanical operations are carried out by the local personnel. The astronomer who gets the observation time just gives instructions and then downloads the data. The Large Synoptic Survey Telescope (LSST), to be built on a dry, cold mountaintop in the Chilean Andes, is expected to be imaging the sky with a 3-billion-pixel digital camera by 2022. And all that data—terabytes of data every night—will instantly be accessible to all kinds of scientists all over the world.

However, even when using such fantastic equipment, astronomers usually work on their own and examine very specific questions relative to their research focus. Once the data is acquired it is rarely used again, although it is usually made publicly available after a certain time. But who wants to bother with the problems of competing research groups and plowing the same field over again?

Automated telescopes do, however, point to another, probably more important, innovation. For example, 10 percent of the Hubble telescope's observation time was allocated for a seemingly pointless endeavor—recording a dark field with almost no known stars and galaxies. The telescope took a picture with every orbit, with light from the dark field arriving at a rate of one photon per minute (which would be comparable to trying to see a light bulb on the moon). The single pictures, each of them almost completely dark, were overlaid and went through image-processing software. The collective outcome of the images was a spectacular picture of galaxies several billion light years away.

Many scientists were enthusiastic about these *Hubble Deep Field* or *Ultra Deep Field* images. In a subtle way, they started a revolution in astronomy. People could use data without being forced to rely on any interpretation. So-called surveys, where the sky is photographed without reference to a concrete question and the data is made accessible to the general public, are becoming more and more important. An outstanding example of this is the Sloan Digital Sky Survey (SDSS), which was conducted with a relatively small but completely automated telescope in New Mexico. The Sky Survey covered half of the northern sky, and a similar project for the Southern Hemisphere is on its way. After an analysis of the raw data, with several calibrations and corrections, there are now nearly 500 million objects on the Internet—positions, red shifts, luminosities, color spectra, and images

Fig. 3. The Milky Way recorded at several different wave lengths, from top to bottom: long radio, atomic hydrogen, short radio, molecular hydrogen, infrared, middle infrared, near infrared, optical, X-ray, gamma-ray.

of stars, galaxies, quasars, and even supernova explosions, which are of particular cosmological interest. For every astrophysical question, this is a true data paradise.

Another survey called COSMOS—the Cosmological Evolution Survey—is a collaboration of about a hundred astronomers, and this survey combines the best images of space telescopes such as the Hubble with ground telescopes to create images of the sky that will cover the entire electromagnetic spectrum. Fantastic!

THE DEMOCRATIZATION OF ASTRONOMY

Astronomers today have less and less need for their predecessors' laboriously learned craftsmanship. Some might feel a little sorrowful about this, or they might recall a quotation from Einstein. "Everybody should be ashamed who uses the wonders of science and engineering without thinking and having mentally realized not more of it than a cow realizes of the botany of the plants which it eats with pleasure." But Einstein would surely not reproach today's astronomers, because the diversification of their professions as "data miners" and "interpreters" opens a new era he probably would have enjoyed. Someone has to process all that raw data from telescopes, and thorough documentation and analysis of the data make it a lot easier for other scientists to attend to the big questions of the universe.

Today, a considerable amount of astronomical precision data is freely available. For instance, you can download the orbits of all known celestial bodies of our solar system from NASA's HORIZONS website.[1] Observations of planetary motions, from centuries ago up until the latest space probes, are all part of the data. Who could ever analyze all this information alone? The International Earth Rotation Service (IERS.org) provides data from radio telescopes on Earth's rotation, and the server of the German Research Center for Geosciences hosts the measurements of dozens of ultraprecise gravimeters around the globe.[2] All of this is just a few mouse clicks away.

The open access to knowledge about the universe can hardly be overrated in scientific progress. Just try to get observation time at a prestigious telescope if you are a researcher at a small, obscure university! That's not to say that getting research time is impossible, but your project will be competing with higher-profile groups who themselves jealously guard their telescope access. And it will be even harder if you work aside the mainstream by testing a hypothesis that is not blessed with "general acceptance."

Galileo himself worked outside the tenets of mainstream beliefs. Although his telescope also brought him trouble, the truth did come out eventually, which is a reassuring perspective! But it is surely one of the most important tasks to create a permanent open access to this treasure trove of astrophysics.

A very different question is, however, how well we understand the vast volume of recent data. Though competition is sometimes useful, doing physics is not a race for the most dramatic supernovae, the most accurate spectrum, or the remotest quasar. We have to actually interpret and understand what all that data is telling us, and we might well be overburdened in evaluating the discoveries that were granted to our generation.

Chapter 3

A SPEEDY REVOLUTION

WHY COSMOLOGY IS GOING THE WRONG WAY

It was early October in 1923 at the Mount Wilson Observatory in the mountains near Pasadena, California. Astronomer Edwin Hubble was contemplating a photograph of the Andromeda galaxy. Eventually, his hand started to twitch. He crossed out the "N" for nova he'd already written on it and jotted down "VAR!" This had to be one of the most exciting moments of his life. He had just identified a peculiar type of star that allowed him to determine the distance to our neighboring galaxy (2 million light years). He had, for the first time, identified a celestial object outside the Milky Way, thus ending the great debate in the astronomy community about whether fuzzy spots on the sky such as Andromeda couldn't also be objects in our own galaxy. Immanuel Kant, who had regarded these nebulae as "world islands" similar to our Milky Way, ended up being right.

Hubble's first big discovery was evidence that our galaxy was not the only one in the universe, and he followed it up with a discovery that allows astronomers to talk about the age of the universe. When an object like a galaxy is moving away from us, the light it emits shifts to the red. What Hubble discovered is that this redshift of galaxy light was related to their distance, which meant that the faster the remote worlds seemed to run away from us, the more distant they were. Obviously, the universe itself was expanding. A neat analogy for this expansion is the surface of an inflating balloon where you have marked dots (each representing a galaxy) with a pen. As the balloon expands, the dots, viewed from each other, begin to recede.

To measure the expansion of the universe, you divide the speed by the distance to get an "inverse" time, the so-called Hubble constant. You can now estimate the time it will take for all other galaxies to double their distance from us, or rewind the "movie" to the point in time at which all galaxies were roughly in the same place: the Big Bang. This time span of 14 billion years is a crude approximation of our universe's age, but it is still considered valid today.

"TREMENDOUSLY FAR AWAY" BECOMES MEASURABLE

The distances we encounter in the universe may be measurable by our instruments. However, they are so large as to be barely imaginable for our minds. Astronomers have been struggling for centuries to make these length specifications trustworthy by using a neat system of intertwined observations, which they call the "distance ladder."

Let's see how to do the first step: stretch out one of your arms in front of you, look at your thumb, and alternately close your left and your right eye. You will notice that the thumb switches between certain angles with respect to the background. If you now measure this angle and the distance your eyes are from each other, you will be able to calculate the length of your arm to about one meter, which is nothing particularly surprising. Now look up at a star and replace the "eye-switching" with spring and fall, which on Earth's orbit corresponds to an eye-distance of 200 million miles. Over that range, the relatively close stars do indeed change their angular position, revealing to us their distance.

This is called a parallax measurement. It was first made up by the German astronomer Friedrich Wilhelm Bessel in 1838, and has been developed to the highest precision by a satellite of the European Space Agency (ESA). Its name, HIPPARCOS, stands for *HIgh Precision PArallax COllecting Satellite*, given in honor of the ancient astronomer Hipparchus from Nikaea, who calculated the Moon's distance from Earth using the parallax method.

By the way, astronomers use a parsec (*pc*) as a unit of distance (which sounds alien to physicists); it is equivalent to the parallax of an arc second and amounts to 3.26 light years. HIPPARCOS' successor, the GAIA spacecraft, will soon provide astronomers with still more accurate measurements of positions for about a billion stars in the Milky Way galaxy. The maximum distance we can measure using stellar parallax right now lies at about 1,000 light years. So, how do we go about measuring even farther?

NEARBY CANDLE OR DISTANT FIRE?

There is a serious and fundamental problem with measuring astronomical distances that go beyond what can be determined using stellar parallax.

A star's visible relative brightness doesn't tell us much about how far away it is. How do we know if a star we're seeing is small and close and glowing weakly, or whether it is really very large, far away, and very bright? To figure that out, we have to know either the distance or the luminosity or the size of the star. We know that an object's brightness lessens by a factor of 100 if we observe it from ten times the distance. However, this helps to no degree if its luminosity isn't known. To deal with this problem, astronomers developed a stable of stellar objects with known luminosity, fondly called "standard candles."

Hubble recognized a "standard candle" called a Cepheid variable star when he scribbled "VAR" for variability on the photographic plate. Just 20 years earlier, Henrietta Leavitt, a reclusive young woman with an obsessive eye for detail, was examining photographic plates at the Harvard Observatory looking for stars with variable brightness in the Small Magellanic Cloud, a companion galaxy of the Milky Way. Her painstaking work, for which she was paid all of 30 cents an hour, eventually resulted in one of the cornerstones of astronomic distance measurements. Leavitt found that the period between two brightness peaks was directly related to the luminosity of Cepheid stars. Just like a larger drum will oscillate more slowly than a smaller one will, a huge Cepheid will oscillate with a period of up to 30 days, while small Cepheids with low light power may change their luminosity on a daily basis.

The important thing is that this oscillating period gives us an independent source for the star's size; thus, the ambiguity between being really bright or being just close vanishes. The striking correspondence of Cepheids revealed by Leavitt became a yardstick with which astronomers could reliably measure the first extragalactic distances. Everything else depends on this second step of the "distance ladder."

A SENSATION, BUT STILL PRETTY WRONG

Hubble had identified a Cepheid in Andromeda, and because he could calculate its distance based on its period and luminosity, he knew it was not in the Milky Way galaxy but much, much farther away. For objects still more distant than Andromeda, Hubble had to rely on other methods and hence greatly underestimated most distances.

Most mistakes in astronomy tend to occur this way. If you look for a star specimen, you will obviously discover their more brilliant exemplars more easily and arrive at an inequitable selection bias in which smaller stars will have been omitted. Say, for instance, you were selecting apparently random people from a crowd in order to determine the average body height. You will likely arrive at a mistaken conclusion, because larger people will have been more noticeable and will have skewed your choices toward taller people. This effect is known in astronomy circles as the Malmquist bias. It plagues astronomers to this day and has rendered many statistical analyses void.

Hubble made another rather subtle error. A Cepheid's oscillating period is not only subject to its magnitude, it also depends on the abundance of the elements heavier then helium. (Astronomers, with an insouciant disregard for elementary chemistry, call all of them "metals.") Heavy elements increase the capacity of a star to store heat. In other words, a star with a lot of "metal" will shine brighter at the same oscillating period. Hubble found such a star in Andromeda and then compared it with low-metal Cepheids in the Small Magellanic Cloud, which led him to underestimate Andromeda's distance. This was merely bad luck, but should also be a warning of the kind of snares that lurk in astronomy.

Hubble would surely find great pleasure in knowing that our most accurate measurement of the Hubble constant is based upon Cepheids in the Virgo and Fornax galaxy clusters, with the data on them all acquired using the Hubble telescope. What could be a better memorial? Hubble died just before the Nobel committee could award the 1953 prize to him.

In 1944, Walter Baade, a German astronomer, recognized Hubble's mistake with the Andromeda distance. Curiously enough, Baade was allowed to use the world's best telescope at Mount Wilson. Ultimately one can't blame Hubble for his faulty assumption. It is, however, remarkable that so many astronomers blindly trusted his authority. For quite some time, researchers seemed to be busy bringing the rest of astrophysics in line with Hubble's distances, instead of questioning them.

Technically careful work on its own isn't always sufficient for achieving accuracy. Systematic errors, like Hubble's mistake of ignoring the chemical composition of stars, are especially likely to occur in astronomy. Often you have to arrive at conclusions by indirect evidence, relying on premises that may turn out wrong. If that is the case, your great measuring accuracy is all for nothing.

When Baade's new determination of distances doubled the size of the universe, and people in the astronomy community suddenly convinced themselves of its correctness, the Russian physicist Lev Landau remarked ironically, "Cosmologists are often in error, but never in doubt." Today, the scientific community again believes that the current picture of the universe is more or less correct. We disregard the fact that we have had to change our worldview numerous times over the course of history.

RESIZING THE UNIVERSE YET AGAIN

Baade was not, himself, immune to errors. He also had to face the embarrassment of one of his students, Alan Sandage, correcting his measurements of the Hubble constant. In this case, Baade had mistaken hydrogen clouds for stars and hence underestimated all cosmological distances again. Today's measurements seem more reliable since they are based on Cepheids, which can be detected more than 50 million light years away. At the moment, the favored value of the Hubble constant (H_0) is, put in the astronomer's awkward way, 72 kilometers per second per megaparsec (3.26 million light years). This translates into a universe that is roughly 14 billion years old. Nevertheless, there is no guarantee that this number is free of all the subtle potential mistakes measurements are subject to.

When we compare the distances of the very close Cepheids we measure using stellar parallax to those in the faraway Magellanic Clouds, we have to make guesses about the concentration of heavy elements (metallicity), which could easily be wrong.

Two years prior to his death, the then 82-year-old Sandage could not make it to the cosmology conference of 2008 in St. Petersburg. He did, however, write a touching letter to Rudy Schild, who then read excerpts of it to the conference's participants.

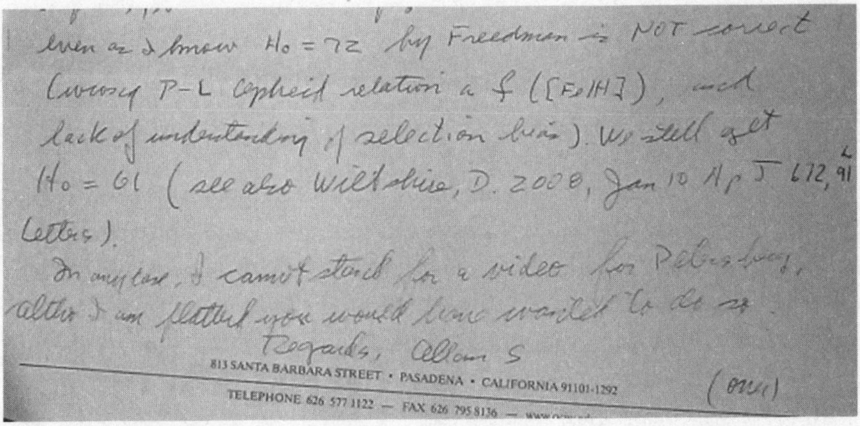

Fig. 4. Excerpt from a letter from Alan Sandage.

"I have abandoned the arena and the game," wrote Sandage, "but $H_0 = 72$ is NOT correct. Wrong Cepheid relation and lack of understanding the selection bias..." So the discussion continues. While Sandage's students insist on a value around 62,* the most recent measurements go as high as 67[1] to 74.[2]

It is hard to measure the Hubble constant precisely, and an honest error margin would be between 5 and 10 percent. And at the very least, this error applies to approximations of the universe's age, even if you will sometimes hear claims of a precision ten times as high. To criticize this might seem like nit-picking, but underestimating errors has become a serious illness in cosmology.

SUPERNOVA: THE CATASTROPHE AS A YARDSTICK

It is impossible to detect single stars billions of light years away. We simply can't "see" that kind of detail. The greater the distance, the greater the difficulties, and cosmology has few methods of reliably measuring distances to remote galaxies. Since the 1980s, we've been getting some help from rare cosmological incidents: supernova explosions, the bright and dramatic deaths of big stars. They can be seen from large distances, but still, they vary considerably in brightness.

Sometimes two stars will orbit each other really closely, and one star continuously draws matter from the other until it collapses and ends in an explosion—a burst caused by overeating, so to speak. Such supernovae, called Type Ia, have similar levels of brightness and have allowed astronomers to determine the Hubble constant at huge distances. The supernovae suggested a universe that was older than what previous measurements had assumed. This was balm to the cosmologists' wounds, who for decades had had to suffer the ridicule of physicists that the cosmologist's model of the universe showed it was, paradoxically, younger than its content.

GLOBULAR CLUSTERS: THE GALAXY'S GRAY HAIR

In order to understand this paradox, we will have to divert a bit and take a look at globular clusters, marvelous conglomerations of stars that the naked eye can only see from the Southern Hemisphere. I had a look at them when I visited an outlying farm in Namibia. The light smog from the capital Windhoek was safely 100 miles away, and the dry desert at an altitude

---

* In units kilometers per second per megaparsec.

Fig. 5. Globular cluster 47 Tucanae, right next to the Small Magellanic Cloud (SMC). Picture taken in Hakos, Namibia, by Stefan Geier.

of 6,000 feet provided ideal conditions for looking out into space. It was a brilliant sight, with the starlit sky featuring a spectacle that you can't imagine when you stay under the diffusely lit palls of smog that hang over our cities.

A globular cluster consists of about 1 million stars far outside the galaxy discs. It is comparable to a suburb suffering from overaging; no more new stars are being born. Astronomers have long learned to categorize stars by their color, from the small, reddish gleaming ones to the very bright white-blue exemplars. (Incidentally, this color has nothing to do with the Doppler shift due to motion, but is a consequence of the surface temperature. The hotter a star is, the bluer it is.) Using the thermal radiation law found by Max Planck around 1900, scientists know that our Sun's temperature is 5,500° C. But how do you figure out the age of globular clusters?

Well, the bigger stars radiate a lot more energy compared to smaller ones and burn up their nuclear fuel much faster. This "fast and furious" life for big, hot stars is exhilarating but short, lasting just a couple of million years. That's why you won't find blue stars in globular clusters. They've long since burned out and died. If there are just reddish dwarf stars the cluster has to be incredibly old, up to 14 billion years according to what we know about the relation of color and brightness. This is a flabbergasting number, even if you assume that spiral galaxies like the Milky Way along with their suburbs were formed right after the Big Bang. A contradiction seems to arise here.

IS THE UNIVERSE AS OLD AS ITS STARS?

Now, you're probably thinking that the idea of 14-billion-year-old ancient stars fits nicely with the value of Hubble's constant, which also corresponds to 14 billion years. But sorry, there's a big problem! We can't assume that the universe always expanded at that actual rate, since gravitation inevitably decelerated the expansion. And if the growth was faster in the past, then the universe must be younger, and we could argue at best by how much (a question that depends on the density and will puzzle us later on). Cosmology was trapped in a contradiction.

But in the 1990s, the supernova hunters made a substantial discovery. They realized that the dim explosions finished more quickly than the bright ones. Once again, the larger ones were acting more slowly, just as it was with the oscillation of the Cepheid stars. The explosion's duration could then be used as a measure of absolute brightness, which dramatically improved the precision when determining the Hubble constant at great distances.

The discoverers of this new cosmic yardstick, Saul Perlmutter, Brian Schmidt, and Adam Riess, were later awarded the 2011 Nobel Prize in Physics. Their conclusions had suggested the greatly preferred older age of the universe. But there was still another conundrum to be solved.

The measurements using Cepheids for the shorter distances didn't fit with the new supernova method for much longer distances. And since distance was a key to calculating the Hubble constant, the inconsistency was obvious, leading to an embarrassing discrepancy by a factor of two between the calculations. The researchers were observing different objects, but no error was found in the measurements of either Cepheids or supernovae. Caught in this predicament, they resolved the issue by deciding that both teams could be right!

That, however, required taking one of the fundamental understandings about the cosmos and flipping it on its head. Rather than a picture where the universe is expanding but gradually *decelerating*, the distance measurements could be harmonized by assuming that expansion of the universe is *accelerating*. This consequence is surely not easy to cope with since it contradicts our experience. It is as if a rock thrown in the sky doesn't slow down and eventually return back to Earth, but rather, falling upward, accelerates away from us.

Anyhow, the paradox of the globular clusters seeming older than the universe was hereby vanquished and, more importantly, the conflict between the competing Cepheid and supernova teams was set aside. Everyone was happy. But it is good to keep in mind that, often enough in history, consensus in the scientific community has been dangerous. It may just whitewash a deeper flaw in our understanding.

Even when the experts all agree, they may well be mistaken.
—*Bertrand Russell, British philosopher*

NEW PARADIGM OR PTOLEMAIC CRISIS?

The supernova data and the new "standard candles" it established rightly led to the 2011 Nobel Prize. Astronomers found assuming an accelerated expansion of the universe to be a comfortable solution, and nobody was particularly alarmed by it.

But of course it raised the question about what the heck could cause this weird acceleration. The answer? Some sort of "dark energy," whose hypothetical properties were designed just for that purpose. However, the term is nothing more than a helpless labeling of something we simply don't understand. Even the proponents of dark energy baldly admit this. Still other theorists are pleased to have a new field (however dark it may be) in which they can do more math. But physics is simple. Another unaccounted-for number—call it dark energy or whatever—is a grave methodological defeat.

There is a historic parallel to the Ptolemaic view of the universe, which prevailed from about 150 A.D. to the Middle Ages, assuming that all motion in the heavens was circular. When astronomers couldn't explain the observed planetary motions with simple circles around the earth, they posited another circular motion on top of the first circular motion, the so-called epicycles. As slightly better observations showed another discrepancy, the centers of the main circles were shifted by a tiny figure called an eccentric. And everything worked smoothly again! Eventually, the whole system collapsed under the weight of all those circles. We could be just trapped in a modern version of such a system.

The science historian Simon Singh wrote in his excellent book *Big Bang* that "every flawed model can be saved by such fiddling around with numbers."[3] The process also recalls a term coined by the English philosopher Owen Barfield, "saving the appearances." It is sometimes simply too hard to throw overboard a model developed over many generations.

Today, everybody would surely agree that embracing such a poorly understood idea as epicycles could just as easily contribute to the erosion of science, the painted-over rust of a crumbling construction, as to the progress of science. No doubt Ptolemaic astronomers found it hard to let their established picture of the universe go. There is no reason to suppose that astronomers and cosmologists today are any less susceptible to wanting to save the appearance of the current model of the universe. Still, given that "dark energy" is really just an ad hoc solution to deal with contradictory measurements, shouldn't we be at least a little bit cautious?

At conferences, however, one notes an irritating euphoria. The accelerated expansion, along with its plethora of fancy theoretical explanations, is brought up wherever astronomers and cosmologists gather. Everybody sees it clearly now! "Otherwise we cannot make sense of the X-ray observations" or "galaxy distributions models require its existence," as a researcher of the University of Nottingham writes with admirable naïveté.[4] However, every scientist should keep in mind that throwing another free parameter or one more number into your model to fix your problems can improve the explanation of *any* data set.

> Cosmological fashions and reputations are made more by acclamation than by genuine scientific debate.
>
> —*Mike Disney, British astronomer*

WHILE FIGHTING THE TIGER AT THE FRONT DOOR, THE WOLF ENTERS THE BACK

Let's recapitulate the scenario one gets by analyzing the supernova data at different redshifts. In the very early universe, the acceleration did not act yet and thus the expansion was slowing down. Eventually dark energy emerged, growing with the expansion until, at a certain moment, the acceleration and deceleration canceled each other out. Would you like to guess when the moment of cancellation occurred? It turns out that after 14 billion years, we have arrived at that point exactly today. Are you surprised? Justifiably so.

Well, that's just plain coincidence, cosmologists reassure us. But this coincidence problem highlights the paradox of the current model. The periods of acceleration and deceleration of the expansion of the universe could just as well be distributed in a very different way, a way that wouldn't make you think that our momentary expansion rate, the Hubble constant, has anything to do with the age of the universe. It is as if you are on a road trip driving with greatly varying speeds. At any given time, you can determine the average speed for a route you have already traveled, but the odds are fairly small that you are, at this moment, driving at exactly that speed.

Dark energy has solved the controversy about the age of the universe, but the coincidence problem ought to make us suspicious that this mending was too easy. The cosmologist Lawrence Krauss has pointed out another funny consequence of this. As a result of the accelerated expansion, some day remote objects will disappear from our visible horizon. For instance, in later eras it will be impossible to observe cosmic microwave background. Future cosmologists will encounter a rather boring universe, so we humans are really fortunate that our brilliant minds evolved just in time for making sense of these unique observations before they disappear from view. Any doubts about being that lucky?

I'll never make that mistake again, reading the experts' opinions.

—*Richard Feynman*

POLITICAL COSMOLOGY: SCIENCE BY A MAJORITY DECISION

A conclusion is the place where you got tired thinking.

—*Martin H. Fischer, American author*

As much as most cosmologists seem excited that everything fits and we have a "safe model," there are few introspective voices, such as that of Bruno Leibundgut, a member of one of the supernova groups that shared a Nobel Prize.

At a conference about dark energy in Munich in 2008, Leibundgut said that it was a nice game to collect up to 1,500 supernovae a year, but he was concerned about the puzzling scatter in the brightness of the explosions. "Just take the best ones," an audience member exclaimed. "This is a dangerous game," Leibundgut replied, "and it's not enough to plot the brightness variation in some diagram. We ought to understand the mechanisms!" All he received were shrugs.

Historical reflections are an unfamiliar virtue to a lot of scientists today. Sometimes the conferences seem like big parties where people enjoy their shared feelings about "consensus cosmology." But what does this mean? Lev Landau's warning that cosmologists are "often in error, but never in doubt" inevitably comes to mind, and some poke fun at the "consensus." Rudy Schild, whom I talked to in St. Petersburg during a coffee break, said, "Which consensus? Do you know who consented? A bunch of guys at Princeton who drink too much tea together." He grinned mischievously as he said this, but it is definitely true that the new picture of an accelerating universe forced its way through the science community very rapidly. All it took was two research teams agreeing and an article in *Nature* ... a speedy revolution.

Today we have a "simple" (at least that's what you read) "standard model" of cosmology. Everyone interprets the data using it, although the ingredients such as dark matter and dark energy are nothing more than fancy names for substances we do not understand—if they exist at all. But throughout the history of science, it has always been much easier to add a new concept than to overturn the existing construction—simply because it's easier to find people agreeing with you. Thus, barely anyone is willing to question the entire system of assumptions that has peaked with the introduction of dark energy. But what is considered progress today may turn out as an erroneous complication. All indications are that modern cosmology has not learned a thing from the example of the medieval epicycles.

Part II

CROSSROADS

Chapter 4

THE BASIC STORY

WHAT EINSTEIN TOLD US ABOUT
GRAVITY AND SPACE-TIME

According to the "standard" or "concordance" model of cosmology, the universe is made up of 4 percent usual matter, while the rest consists of invisible substances such as dark matter and dark energy. The vast majority of astronomers consider this model to be an essentially correct description of the cosmos.

However, one gives the concordance model too much credit by viewing it as a unique consequence of Einstein's celebrated theory of general relativity found in 1915. All of its equations relevant to cosmology can be derived from Newton's laws as well. For the small accelerations that govern cosmological dynamics, Einstein's theory perfectly merges into Newton's law for weak gravitational fields.

The same holds for Newton's laws of motion and the theory of special relativity that Einstein had come up with already in 1905. Relativity builds upon a surprisingly simple insight—the speed of light is always the same, even if one views it from a moving system like a car or a rocket. Galileo was the first to recognize such a relativity principle, and he tried (largely in vain) to explain to the geocentric folks of that time that cows and trees won't be flung into space if Earth is conceived as moving. As a matter of fact, there is no physical experiment whatsoever that you can conduct that will tell you whether you are at rest or you are moving in uniform motion with everything around you.

NOTHING BUT PURE LOGIC—MOVING CLOCKS RUN SLOWER

All Einstein did was take Galileo's principle seriously and apply it to a modern experiment—the measurement of the speed of light. If you are speeding down the highway and you switch on the headlights, the light travels at the speed of light, no matter how fast your car is going. It travels at the speed of light whether you observe it sitting in the moving car or standing on the roadside!

Einstein resolved this seeming contradiction by discovering that time is measured differently in a moving system than in one at rest, with the strange result that clocks in motion have to run slower. As startling as this may seem, the famous factor by which time passes slower (the square root of $1 - v^2/c^2$, where v is velocity and c is the speed of light) is derived by simple logic.

The name of this surprising fact is "time dilation." It is present in our everyday life, but the effect is so small that it can hardly be seen. By flying from Munich in Germany to Canada to visit my coauthor Sheilla in Winnipeg, I would gain all of about ten nanoseconds!

There are plenty of fail-safe measurements of the effects of relativity. Deduced from an elementary principle, they are also completely consistent throughout whatever thought experiment you might conduct. Thus, do not let yourself be impressed by some crackpots on the Internet who polemicize against Einstein and show you an "apparent" flaw in his theory by using a five-sentence argument. Unfortunately, Einstein is one of those founding fathers like Freud or Darwin, who magically attract peeing dogs.

There aren't the slightest doubts about the theory of special relativity. All of its predictions, such as high speeds leading to time dilation, or mass increase conveyed in the famous equation $E = mc^2$, are in excellent agreement with the observations.

A 17-YEAR-OLD IS SURFING THE LIGHT WAVE

I have no talents. I am just passionately curious.

—*Albert Einstein*

Einstein's way of thinking should, by the way, make us reflect a little. When attending high school in the Swiss city Aarau, he imagined what would happen if he was moving parallel to a light ray. The basic idea of relativity had been circling in his head since then. This demonstrates that the ability to marvel and ask the right questions is much more important than being a high-flyer in math.* Einstein approached the generalization of Newton's theory of gravitation in a similar fashion.

---

*He wrote: "I was lucky in encountering books which did not insist too much on mathematical rigor…"

Imagine you are somewhere in weightless space, in a windowless room that is constantly being accelerated by a jet engine. If the thrust is strong enough, your body's inertia will push you to the floor, just like it would in a windowless dungeon on Earth under the force of gravity. This is another case where there is no way to experimentally distinguish one situation from the other.

This thought experiment is intimately related to the nature of mass. Even without a gravitational field, we feel the *inertial* mass of a body as something that opposes acceleration. On the other hand, *heavy* mass is perceived as weight in a gravitational field. Are these two masses really the same? For Einstein this was "a thing I wondered exceptionally about,"[1] leading him to the conclusion that, yes, both masses are the same. And he built his entire theory upon this equivalence of accelerated inertial mass and heavy mass that gravitates.

Acceleration causes velocity, which influences, as we have seen, the duration of time. Thus, if accelerations have the same effect that gravitational fields do, Einstein was reasoning, these fields necessarily affect our clocks as well, even if we are just hanging about. Time dilation is therefore a key effect of the general theory of relativity as well.

Imagination is more important than knowledge.

—*Albert Einstein*

NATURE'S TIMETABLE: FIXED BY SPECIAL AND GENERAL RELATIVITY

Again, it's an easy exercise to calculate the factor by which time passes slower. For general relativity, the expression $1 - v^2/c^2$ from special relativity is replaced by $1 - 2GM/rc^2$, where G is the gravitational constant, M is the mass of Earth, and r is the distance to Earth's center.

Accordingly, time runs slower for one day in New York than at the top of one of Colorado's many mountains. But the difference is a mere forty nanoseconds. That's not quite enough to make New York more relaxing, but it is indeed measurable with atomic clocks.

There is another interesting example that neatly illustrates how time passes in special and general relativity. Imagine throwing a stone with a clock attached to it into the air. Your goal is to make this clock go as slowly as possible to maximize time dilation. According to the special theory of relativity, due to its takeoff speed, the clock will already be ticking somewhat slower. However, if the stone were to continue flying upward without constraint, it would enter a weaker gravitational field, making the clock tick faster again. Thus the rock will stop, start falling with increasing speed (slowing down the clock) to once again enter a strong gravitational field (same effect). Now, what is the optimal timetable to have the time of the stone's clock elapse the least? Precisely the path that results from the "conventional" laws of motion for the trajectory of a rock! If you are a physicist, this makes you happy.

No wonder that principles such as that of least time, developed and put to good use by the mathematicians Euler and Lagrange, are considered benchmarks for finding fundamental laws of nature. And, of course, there is a crucial role for energy when we are talking about time and motion. It turns out that the sum of kinetic, potential, and other forms of energy are conserved at all times and during any event. It doesn't matter whether it is the swinging of pendulums, the oscillation of springs, or the falling of a stone. The concept of energy is a remarkably powerful way to help us analyze processes regardless of the details of motion and time of the moment.

However, for this very reason, energy conservation is basically a rule grown out of the desire to have time-independent laws of nature. But *are* they really time independent, if we consider cosmological periods? Don't forget that all observations, astronomically speaking, have taken place in the very short period of time that *Homo sapiens* has been doing physics. For such short time spans, energy conservation is an excellent fit. However, a tiny variation in the flow of time would by no means be eye-catching on the cosmological scale. Therefore, we have to scrutinize the entire concept of energy when talking about cosmology and see if Nature's timetable has really been the same for 14 billion years.

MACH VS. NEWTON: SPACE WITHOUT MATTER DOESN'T MATTER

We can see already that space and time are not such simple concepts, especially when we are considering the cosmos as a whole. It is one thing to have increasingly sophisticated technology for ever more precise clocks to measure time and spacecraft to measure distance. But it is the very nature of this basis for our perception that is still puzzling. What is time? What is space?

There is hardly anyone who has contemplated the concept of time more deeply than the British physicist Julian Barbour. His argument—that there is no such thing as time—seemed pretty absurd to me at first. And when I heard that he had devoted 35 years to this idea, I thought it was an excuse to justify all the time he had frittered away. However, there's more to it.

When we say that one second today is the same as one second yesterday, what does that mean? This is *not* a play on words, since all we have as measures of time is the observation of Nature's periodicities. Envisioning an absolute time, which flows without any relation to matter, might be completely false, as false as Newton's notion that absolute space without matter exists.

In a famous thought experiment, Newton claimed that in a rotating bucket filled with water, the centrifugal force will make the water level rise at the inside wall of the bucket. In contrast, if the water level stays flat, it means there are no forces acting on the water. Such an absence of force and acceleration, according to Newton, defines "absolute" static space.

However, in 1887, the physicist and philosopher Ernst Mach from Vienna criticized this interpretation of the rotating bucket with a profound objection. "Nobody can say how the experiment would turn out, if the bucket's walls became increasingly thicker and more massive—eventually being several miles thick." Mach suggested that in such a case, the centrifugal forces may vanish. He argued that there is no absolute space but rather that it is distant celestial bodies that tell us what means to be at rest. In other words, all that matters is our motion relative to other masses out in the universe, without presupposing an inertial framework of absolute space. This is called *Mach's principle*.

NO MATTER, NO SPACE? NO MOTION, NO TIME?

Julian Barbour has written books and organized conferences on Mach's principle. Barbour's central idea, portrayed in his book *The End of Time*, is that time is defined through the various periodicities we observe in Nature. It is a profound generalization of Mach's principle. Barbour is a truly unconventional thinker. His theory, which even calls into question the expansion of the universe, is so far off the mainstream that cosmologists must fear for their jobs if it turns out to be right.

Since I had pondered Mach's principle for quite a while myself, I was curious as to what such an original researcher had to say about it. Therefore, I was delighted when Barbour agreed to meet during a conference near Cologne in 2008. After he got his PhD in astrophysics, he wanted to focus on fundamental physics. However, this was not what the quickly changing academic world had to offer him. Instead, he made his living by translating Russian physics journals on a farm in southern England, while he quietly developed his ideas. Barbour is the rarest of creatures, a freelance theoretical physicist.

When we talked, Barbour laid out his ideas calmly, without the missionary zeal so often found in unorthodox scientists. And he also listened to me when I told him about the ideas of Robert Dicke, an American astrophysicist who also had struggled with Mach's principle. Barbour's entire nature is one of modesty. (He wears a jacket with zippers "too tight for pick-pockets," as he amusingly accounts.) In the scientific community he is an "outlaw," but I think physics would be in a better shape if there were more researchers like him. We kept up our discussion for several hours while I accompanied him to the airport. His entire hand luggage was a book.

WAS YOUR WRISTWATCH REALLY UNIMPRESSED BY THE BIG BANG?

What can we learn from Barbour? For one thing, we can recognize that it is probably much too naïve to think that time is something "objective" that runs independently from what happens in the rest of the universe. Imagine a wristwatch ticking away from the beginning of the universe, telling us when the Big Bang took place, when the atomic nuclei formed, and a little later, when cosmic background radiation emerged. But unless atoms exist, there is nothing to tick. There really is no way to count time from the precise moment of the Big Bang.

For this simple reason, we should remain skeptical about the fairy-tale stories about what happened in the 10^{-35} seconds after the Big Bang. No clock can measure such tiny intervals, and although this is evident, many cling to this all-too-simplistic picture of time.

Unfortunately, once you discard the idea of an imaginary wristwatch ticking away time from the moment of the Big Bang, trying to define time is a lot like trying to nail Jell-O to the wall. The cosmologist John Barrow has noted, "The question if there is a unique absolute standard of time which globally is defined by the inner geometry of the universe, is a big unresolved problem of cosmology."[2] And it is not an unimportant one.

CURVATURE, DISTORTED SPACE-TIME, AND SHORTEST PATHS

Einstein's theory of general relativity doesn't explain time either, but it provides some insight into the variability of time and length scales, which leads directly to geometry. Gravity is directly related to curvature of space. The shortest path connecting two points such as Rome and New York isn't a line in an East-West direction, but a so-called geodesic that crosses the Northern Atlantic, as every pilot will know.

The best known geodesics, or great circles, on Earth are the equator and the longitudinal circles going through both poles, but there are many others.* However, it is impossible to draw rectangles on a sphere using these geodesics. The lines will never fit—try for yourself.

One can now imagine time as a fourth dimension alongside the three spatial ones, whereby clocks take the role of the rulers that measure temporal distances. But time runs slower in gravitational fields; thus, if you try to draw a "space-time rectangle," the time side will fail to close it! This distortion of time and length scales in general relativity is therefore a nice example for curved geometry, a fascinating property of nature.

STROLLING AROUND THE GLOBE ROTATES YOUR HEAD

This may be of interest only to people who get a kick out of geometry, but Italian mathematicians such as Tullio Levi-Civita developed an interesting method for describing curvature as early as the nineteenth century. Imagine walking around the globe while keeping an arrow (a vector) at the same horizontal direction and always parallel to the Earth's surface. Now continue hiking on the same degree of longitude south from the North Pole to the equator, then westward along the equator, and then back north to the pole.**

You will realize that the arrow isn't pointing in the initial direction anymore, even if you've put all your effort into keeping the arrow fixed. This is the astonishing consequence of walking along a curved surface. This idea of vector transport, called "connection," is an important concept in general relativity, which, probably to your relief, won't be examined here any further.† But you have already got a glimpse of the field in which Einstein worked on his unified theory.

EINSTEIN, CARTAN, BIG AND SLEEPY HEADS

Mathematical tools such as differential geometry were anything but easy for the young Einstein. In his memoir he wrote that "profound knowledge of the basic principles of physics is tied up with the most intricate mathematical methods. This dawned upon me only gradually after years of independent scientific research."

---

* Google Earth displays geodesics when you mark the path between two points on the globe. It becomes funny when you choose nearly antipodal points: the path will tumble around crazily.

** Instead of taking this rather long walk, try to do it with the coconut as described in chapter 1.

† However if you are interested, the result of the twist can be obtained by summing up the curvature of the surface surrounded by the path. This interrelation of curvature and differential geometric connection is completely analogous to the theorems of Gauss and Stokes in electrodynamics and the reason differential geometry has attracted many physicists.

So in 1922, Einstein didn't yet comprehend an idea that he later considered the royal road to a unification of gravitation and electrodynamics. The French mathematician Ellie Cartan had told him about a subtle change of the rule for transporting a vector around a sphere: just keep it parallel to circles of constant latitude. This is also called the "Columbus connection," always westward! By just perceiving the geometry of the globe in another way, gravitation isn't described via curvature, but through torsion*—a similar, albeit not as intuitive, concept. Years later, Einstein rediscovered Cartan's idea, and a highly interesting correspondence emerged.[3]

Incidentally, today most physicists regard this notion as a dead end. A gravitational physics professor, the chair of his department, once told me, "In his last thirty years, Einstein just did nonsense." Such an...let's put it gently...ambivalent stance toward Einstein's work is widespread. On the one hand, his established theories are seen as a holy book, and changing one iota of their formalism is considered a heresy. On the other hand, his creative wrong paths, inseparable from his greatest achievements, are considered to be old-fashioned hokum.

Einstein's later work is easy to appreciate for every native German, but, for example, the Session Reports of the Prussian Academy of Sciences, in which Einstein published most of his work after 1914, still haven't been translated into English. My friend José Vargas from Spain, who worked at the University of Columbia in South Carolina, angrily wrote to me more than ten years ago, "The Russians, despite their incompetent system, managed to translate it a long time ago!"

There is, however, the Einstein Papers Project at Princeton University. It consists of a considerable number of scientists and historians flipping over every single letter in Einstein's notebooks and meticulously interpreting them. The whole team needs five times as long for close reading as Einstein needed to write his own notes down. I am sure they are being paid until they're finished. At the current tempo, they should be done by 2060.

WEIRD THINGS ABOUT ACCELERATION

Let's take a look at the content of Einstein's notebooks. He explained his relentless search for a unified field theory quite plainly. Since gravitation can be described through geometry, he suspected, one ought to search there, too, for concepts reflecting the laws of electrodynamics. Indeed, the electric force acting on charged particles has a notable resemblance to the gravitational force acting on masses in Newton's law. When you double

---

* Torsion, unlike curvature, could encompass electromagnetism, as Einstein conjectured.

the distance of the objects carrying mass or charge, the force is reduced by one-fourth. The mathematical structure, called the inverse square law for distances, is the same.

It is quite intelligible that Einstein set out to describe this simple formula of electrodynamics in a more general framework. However, classical electrodynamics has its own problems. One is that steadily accelerated electrical charges radiate energy. But remember that, due to the equivalence of inertia and weight, acceleration and gravity are fundamentally the same thing; thus, charges should radiate energy in a gravitational field even when they just sit there. This remains an unresolved puzzle.

Actually there are far worse problems arising from the fundamental law that accelerated charges radiate light.* You may think that once the acceleration is known, physics has a formula for calculating the amount of radiation. Unfortunately, it doesn't, as Richard Feynman explains in his *Lectures*.[4] Feynman's books refreshingly differ from many others in that they address unsolved problems, rather than camouflaging them under a bunch of brilliant mathematical formulae.

The deeper reason for the mystery of the inability to calculate radiation is that classical electrodynamics is inconsistent. If you combine the formula for energy density with that of a force field, a single electron has an infinite amount of energy, and due to Einstein's $E=mc^2$, it also has an infinitely great mass. Something has got to be *very* wrong! And if people tell you that quantum electrodynamics fixed the problem, don't believe it. Feynman, who got the Nobel Prize in 1965 for his role in developing quantum electrodynamics, says it does not.

WHERE GRAVITATION BEATS ELECTRODYNAMICS

Researchers have been able to observe charges at extremely high accelerations. Laser beams pulsed in the femtosecond range (10^{-15} seconds) host unbelievable energy densities. To get an idea of what this means, consider that such an ultra-high-peak power laser needs the equivalent of 50 nuclear power plants as an energy source, but only for a few femtoseconds a time.

The way electrons radiate once they are accelerated by the enormous electromagnetic field of such a laser was described in an article in *Nature* some time ago.[5] I wondered then how the authors had managed to calculate the radiation accurately, and asked them in an e-mail about the problem discussed in Feynman's textbook. After a while, a member of the group replied very politely that he didn't see any difficulties in calculating the radiation,

---

* We say "light" here for electromagnatic radiation of any wavelength.

but later admitted, "At this point, we do not understand the experiment." Fine, but who should? And why didn't they tell the readers of *Nature*?

Unfortunately, a neat result—especially one using the latest high-tech gadgetry—is always more fashionable than grappling with fundamental problems. Meanwhile, this article has been quoted nearly a hundred times, but to date nobody knows precisely how strongly accelerated electrons radiate light.

To add insult to injury, the elegance of the equations of electromagnetism Maxwell developed in 1865 is not enough to formulate the theory. One still needs the force law discovered by the great Dutch physicist Hendrik Lorentz, which tells a charged particle how to move in an electromagnetic field. The two pieces are stitched together for the theory of electrodynamics.

By comparison, general relativity is a polished thing, consisting of Einstein's equations and nothing else. Spacetime tells matter how to move; matter tells spacetime how to curve, as John Wheeler, a pioneer of gravitational physics, once succinctly expressed it. For a long time after its development in 1915, the theory of general relativity was based upon meager observational evidence. However, starting in the 1960s, it passed a series of precision experiments. They are referred to as "the four classical tests."

EINSTEIN WAS RIGHT, TO THIS DAY

There was a spectacular observation not long after the theory of general relativity was published. The deflection of light rays through the sun's gravitational field—the bending of light due to the curvature of space—was confirmed in 1919 by Arthur Eddington's legendary solar eclipse expedition, which put Einstein in the *New York Times* headlines. Newton, whose theory could not explain the observation, now had a successor.

At the time, Einstein himself was already convinced of the correctness of his theory. After all, it could explain a tiny anomaly of the planet Mercury, which had baffled astronomers since 1859. Every century, Mercury's perihelion, the elliptical orbit's point nearest to the Sun, inexplicably shifted by 43 arc seconds. As Einstein wrote in his memoir, when he first performed the crucial calculation to explain the anomaly, he couldn't sleep because his heart was throbbing.

Other predictions took a little longer to be verified. The gravitational redshift of atomic spectral lines is quite easy to understand. Light has to lose energy as it moves away from a gravitating body. It cannot slow down because the speed of light is constant, and therefore, it reduces its oscillation

Fig. 6. Looking out of Einstein's apartment today. Here, in Kramgasse 49 in Bern, he wrote down his special theory of relativity in 1905. The distinctive "Zytglogge" (time bell) in the background may have inspired his thoughts on the nature of time.

frequency. This leads to the increase in wavelength visible as a shift to the red part of the spectrum. That effect was measured in Earth's gravitational field in 1959, the first precision test of general relativity. And it was a terrific match, honored by a Nobel Prize for the experimenters.

In the 1960's, Irwin I. Shapiro from MIT discovered another test that had gone unnoticed. That test became possible with the first radar ranging measurements of the planet Venus, so it is quite understandable that Einstein hadn't thought of this experiment. Because both time and length scales contract in gravitational fields, the signal, traveling from Earth to Venus and back, is delayed whenever the radar beam passes the sun closely. This, too, was splendidly verified.

And last but not least, Einstein's insight that inertial and heavy mass are the same withstands the most accurate experiments. Maybe for that reason, people don't feel any more the necessity to muse about questions like the origin of time. It was, however, Einstein's way of thinking.

BINARY STARS: EXTREME GRAVITATIONAL FIELDS

Some years ago, my students and I had the opportunity to look at the stars with a fairly good amateur telescope far away from the haze of the city of

Munich. We had been invited by a young astronomy student I had met at the Contest for Young Scientists, where he later went on to win the second place in Europe.

He showed us galaxies and globular clusters, as well as his self-built spectrograph, with which he had measured the finest color differences in a binary star system. Two stars circulating each other is not uncommon, but these two were so close to each other that their light merged into a single point. An analysis of the light's wavelength can prove, though, that there are indeed two stars. At times one finds a relatively coherent color spectrum, whereas on other days the light is simultaneously shifted to the blue and to the red: one star is approaching us, the other one is moving away!

A binary star system of similar structure had caused excitement in 1974, because, truly a fluke, one of the stars was a pulsar, one of those fast-rotating stellar remnants of supernova explosions. The orbiting time and the light's redshift helped calculate the distance of the stars, and—the real highlight—additionally showed a perihelion shift* of the elliptical orbit in a gigantic magnitude of 4.2 degrees per year, over 35,000 times that of Mercury's. General relativity was apparently still true far out in our galaxy, and this was worth a Nobel Prize for discoverers Joseph Taylor and Russell Hulse in 1993.

GRAVITATIONAL WAVES: ONE STILL CAN'T HEAR THE GRASS GROWING

After the sensational star-weighing, scientists' interest in pulsars surged. That's because, based on the predictions of general relativity, the strong disturbance of space-time curvature around pulsars should radiate great ripples out into space in the form of gravitational waves. And indeed, the Taylor-Hulse pulsar was observed to slow down, as was expected from such a radiation. This was, however, not yet a proof.

More than 30 years have passed since then, and generations of gravitational wave detectors have been developed. Despite immense efforts to register the waves directly, no signal was found that fulfilled the hopes the pulsar had awoken. Such failures make the detection of gravitational waves one of the most frustrating businesses in physics.

A couple of years ago I attended a seminar where the speaker claimed that a variety of objects—supernovae, pulsars, gamma-ray bursts—were about to reveal gravitational waves. In an extensive analysis shortly afterward in

---

* Being overly correct, one may say periastron advance or shift.

2009, nothing was found.[6] A more recent review article promises detection by 2014.[7] We shall wait with patience. As of yet, the pending experimental proof of gravitational waves is hardly viewed as a serious problem for general relativity. Although Einstein's theory has been tested only within our solar system, it is considered to be too "elegant" for astronomical craftspeople to be allowed to cast doubts on.

Nevertheless, one shouldn't forget that the universe is larger by a factor of 10^{14} than our solar system and has 10^{23} times the mass.* Around 1900, people believed that we could apply Newton's laws of motion to atoms. With what we now know about quantum phenomena, we can smile a little at such naïveté. But that was an extrapolation of only ten orders of magnitude. Viewed in this way, believing that general relativity applies to the entire cosmos—an extrapolation of 14 orders of magnitude—is 10,000 times more frivolous.

---

*There is nothing contradictory in talking about the size of the universe, since it simply refers to the region from which light could reach us since the Big Bang, which is called the horizon.

Chapter 5

STILL A MYSTERY

NEWTON'S GRAVITATIONAL CONSTANT: FROM ENGLAND TO THE EDGE OF THE UNIVERSE

Albert Einstein created a wholly new, geometrical concept for gravitation with his theory of general relativity, and he did so by building upon Isaac Newton's gravitational constant, the so-called "big G." G is a constant of Nature that tells us about the universal strength of gravity. Although Newton's law beautifully described the movement of celestial bodies, he could not deduce the value of G from that. It would have required an additional measurement of the gravitational attraction between two small masses, a task Newton himself regarded as hopeless.

It wasn't until a century later in 1798 that Henry Cavendish succeeded in developing an experiment called torsion balance, which is being used almost unaltered to this day to determine G. Cavendish utilized a twisted fiber with a couple of skillfully suspended metal spheres to measure the tiny gravitational force that was a billion times smaller than the weight of the spheres!

This was fantastic, since astronomers could make use of the laboratory value of G to eventually figure out the mass of Earth and the Sun. Since then, the mass of every celestial body—indeed every estimate of mass for the entire universe—is measured relative to the masses of the Earth and our Sun, which in turn depends on the Cavendish experiment. To date, there are few alternative setups yielding a similar accuracy.

A LATE REVERENCE TO GALILEO

Being the balance scales of the universes, these experiments are of great significance, and a couple of years ago I had the opportunity to attend a workshop where the most recent results were being presented. The little conference was held at a wonderful location in the medieval center of Pisa, embedded in the beautiful scenery of northern Italy. As you may know, the town was Galileo Galilei's stage, and where—according to the legend—he conducted his falling bodies experiments from the Leaning Tower of Pisa.

Since the 1980s, measuring the gravitational constant has been considered a business unworthy and too unexciting to justify an international conference. The Cavendish experiment had been continually improved up until then, and everyone seemed happy with a very accurate measurement of G.

However, in 1995, researchers of the German Physikalisch-Technische Bundesanstalt (PTB), which is roughly equivalent to the National Institute of Standards and Technology (NIST) in the United States, shocked experts with a new measurement of "big G." Their value stood out by far from the accepted margin of error. Others might have swept that under the rug, but the group had the courage to publish their observations.[1] The experiment, however, contained a mistake, and when other scientists tried to replicate the G measurements, their results contradicted the high value obtained by the PTB researchers. What was surprising was that the researchers challenging the new result could not reproduce the established figure either. Even worse, their results were all contradicting each other.[2] As a result, CODATA, the official "rating agency" for the constants of Nature, downgraded G's credibility by increasing its error margin by a factor of ten. This was just a bit embarrassing, especially with respect to other constants of Nature, which are all known with an accuracy of many decimal places. The whole story is somehow disquieting. If the fundamental law of gravitation is true, its constant G should always have the same value, whatever method you may use for measuring it.

A VERY FRIENDLY DISPUTE: BUT IS THE CONSTANT A CONSTANT?

Modern versions of the torsion balance, 200 years after Cavendish's efforts, avoid the error-prone force measurement through the fiber's torsion and use subtle electronics with a compensating force. Whenever we replace old-fashioned mechanics with electronics, we gain a pretty high level of precision. Just think of the difference between a gramophone and digital hi-fi.

The new method[3] for measuring *G* was presented at the Pisa workshop by Terry Quinn from the Bureau International des Poids et Mesures (BIPM), as well as from Jens Grundlach from the Eöt-Wash group* in Seattle. Their results had an impressive precision, but there was still an inexplicable gap between them.

The next evening, we had a wonderful conference dinner in the romantic convent building. I sat across from Terry Quinn and asked him why he was so friendly with Grundlach, given their dispute over the discrepancy in the value of *G*. He shrugged and emphasized the difficulties of the experiments, and then, with a little smile, added, "Of course, he's wrong!" But this was rather playful than an attack; they didn't *really* want to hurt each other.

The friendly rivalry between Quinn and Grundlach seemed to me to be partly due to the fact that they didn't completely trust their own accuracy. Grundlach's experiments showed greater accuracy, but weren't as convincing because he had averaged a series of measurements. This is a fairly common bad habit among experimenters, which allows them to give a smaller statistical error margin, which in turn leaves the impression that the experiment is more accurate than it really is.

Also, the CODATA commission, probably concerned about the reputation of their constants of Nature, could not make friends with the claimed accuracy, and retained a margin of error for *G* ten times higher than that of the Eöt-Wash group.

Since that allegedly precise but faulty measurement in the 1980s, experimentalists have gotten more careful. Other methods have been tried like a swinging pendulum, but still today, the latest results contradict each other.[4] Symptomatically, almost all results fall outside of the range of error of the others. Hence, everyone seems to be overestimating their own precision.

NEWTON AND CAVENDISH: HUNDRED YEARS AND A FACTOR OF 10^{20} IN BETWEEN

Newton's early doubts about the possibility of measuring the gravitational constant were caused by the immense mass difference between celestial bodies and the famous falling apple that is supposed to have bopped him on the head. Strictly speaking, we are extrapolating Newton's law over many orders of magnitude, because an independent estimate of Earth's mass via geological considerations is very inaccurate.[5] Therefore, physicists put great

---

* Besides the obvious reference to the state of Washington, the name "Eöt-Wash" helps you pronounce the Hungarian Baron Eötvös, who in 1907 performed his groundbreaking experiments on inertial and gravitational mass.

effort into measuring G with masses greatly exceeding those of an apple or a metal sphere.

There was a very inventive experiment conducted in the Italian Alps not too long ago, in which a high-precision gravimeter was positioned underneath the storage lake of an electricity company.[6] Due to their incredible sensitivity, those instruments could feel the gravity decrease due to changes in the level of the overlying water, where the level variation was readily visible. Overall, however, the method is unfortunately not very precise.

As early as the 1980s, there had been experiments indicating a deviation from Newton's law.[7] Researchers had measured the change in gravity on towers, in mines, and in different ocean depths, and one group had even dug a one-mile-deep hole into Greenland's ice cap in order to place a gravimeter into it.[8] However, there is a subtle difference between the lake experiment and this one: the gravimeter in Greenland's ice was moved and thus the measured distance to differently dense parts of Earth's crust varied— unfortunately, an inevitable source of error. The surprising result from Greenland that had shown a deviation from Newton's law (an "anomaly") was explained by assuming that the rock below the ice might have had a different density. One usually prefers to believe that geologists may have erred about that rather than doubting Newtonian gravity.

To summarize, we cannot conclude yet that there is a flaw in Newton's law. But it is surely not well tested under all aspects, given that there is a very big gap between the mass of a couple of cubic miles of ice and the mass of a celestial body. We need to conduct further measurements of G at different mass scales. It would be dangerous if the research field studying "big G" were to once again fall into a Rip van Winkle sleep.

FUNDAMENTAL CONSTANTS: NATURE'S MYSTERIOUS BUILDING BLOCKS

At the beginning of time, the laws of Nature were probably very different from today.

—Paul Dirac, Nobel laureate 1933

Since we are wondering about G and its mysteries, we should pause for a moment and bring to mind the essential role of the constants of Nature in physics. Unlike numbers such as the boiling temperature of water (212°F) or Earth's local gravity ($g = 9.81 \text{m/s}^2$), we are talking about Nature's signatures that are supposed to be valid for the entire cosmos.

Extraterrestrial civilizations would certainly use different units from the meters, seconds, etc., that we use, but undoubtedly their kids would learn that the speed of light is a fundamental constant. We are, after all, in the same universe.

The most important of this signature numbers are the speed of light c and Planck's constant h. The tiny value of $h = 6.62 \ 10^{-34}$ kgm^2/s was found by Max Planck in his famous theorem for light radiation, but its true importance was revealed by Einstein, who realized—to Planck's opposition—that light quanta carry the energy $E = hf$ (with f being the light frequency). The very existence of light quanta is one of the big riddles of physics.

Fundamental constants are Nature's cryptic messages and therefore, it is a revolutionary leap if we happen to discover a connection between them that allows us to calculate one constant by using the others. These are the great puzzles to be solved. The epoch-making discovery that light is an electromagnetic wave, for instance, is expressed by linking electric and magnetic constants, ε_0 and μ_0, to the speed of light c by means of the formula $1/c^2 = \varepsilon_0 \mu_0$.

That is why the question of whether G might not actually be a universal constant, equally important as c and h, is quite a hot topic. This would suggest that G can be expressed by other quantities in the universe. Yet the somewhat discrepant measurements of G would not seem to justify such a wide-ranging hypothesis on their own.

BLACK HOLES—OBSERVED BY CALCULATIONS

Black holes are commonly regarded as a prediction of general relativity, but interestingly, their existence already follows from Newton's law, which assumes that G is a constant. John Michell, an English natural philosopher, realized as early as 1784 that every gravitating body has a minimum velocity needed for anything to escape from it. If this velocity happens to be above the speed of light (which never can be reached), well, bad luck, you're in a black hole! Don't worry, this doesn't happen often. Our planet would have to contract to the size of a ping-pong ball to become a black hole. For every mass M, however, you can compute the size of a black hole using the Schwarzschild radius $r_s = 2GM/c^2$ (named after a German astronomer). It is the distance from the center of a mass where gravity does not grant an escape for light any more.

The wondrous effects at the edge of the Schwarzschild radius lend it a special aura. Time, according to general relativity, goes by infinitely slowly as you approach the edge, so that no object can ever come back from there.

Any light that managed to leak out would lose its entire energy due to the enormous gravitational redshift and hence, never get out. Whatever is inside the Schwarzschild radius is a hungry dark beast, devouring everything within its reach with mathematical assurance.

There is lots of theorizing about black holes, and mathematicians worry about what might happen to their theorems—since at the edge of a black hole, Nature seems to be dividing by zero, a forbidden operation. Interestingly, black holes do not provide a real test of general relativity because nobody has yet measured the size of a Schwarzschild radius. The first evidence—albeit indirect evidence—for a black hole of about ten solar masses was an X-ray source in the constellation Swan. Astronomers couldn't explain its irregular radiation in any other way.

Moreover, a group at the Max Planck Institute for extraterrestrial physics observed, with extraordinary patience, a couple of stars orbiting the center of the Milky Way at extremely high speeds. An enormous concentration of mass must be there, millions of solar masses at least. Understandably, astronomers interpret this as a "supermassive" black hole located at our galaxy's center, because no other object could be identified there (And still, there are a couple of contradicting observations.[9]) Strictly speaking, however, the evidence for a black hole is partly based on seeing nothing.

WOULD NEWTON BE HAPPY ABOUT ALL OF THIS?

It is somehow hard to imagine that black holes could actually have quite a small density. In principle, a black hole could be made of a pail of sand, if it extended a little farther than Earth's orbit! But black holes cannot be made of normal matter, otherwise their mass would increase by the third power of their size, just like the volume of a sphere. Instead, mass is just proportional to the size of the black hole, which is quite strange. One may wonder, however, if this fact is intimately related to the existence of the speed of light c and the gravitational constant G. Because if you divide c^2 by G, you obtain the units *kilograms per meter*, which is again mass per size. This is already a puzzle, but there is more to come.

While a black hole of a single solar mass would be just a few miles in diameter, the mass of a big galaxy, concentrated into a black hole, would extend to about one light year. However, a black hole with the mass of the universe—and this is where you really have to wonder—would be about as big and dense as the visible universe.

Of course, the first estimates for the (very small) density of the universe were not available prior to Hubble's measurements around 1930. Arthur

Eddington, the astrophysical authority of the time, first noticed that the gravitational constant may be interrelated with the properties of the universe. By dividing c^2 by G as above, we obtain roughly the same value as dividing the universe's mass by its radius. This is quite extraordinary.

MACH'S PRINCIPLE: WHAT'S THE ORIGIN OF THE INERTIA OF MASSES?

There are considerable uncertainties in estimating the mass of the universe, and Eddington's observation was an approximate coincidence, of course. On the other hand, we are talking about such huge numbers that even an agreement in orders of magnitude (multiples of ten) is surprising. The modern name given to the enigmatic cosmic coincidence found by Eddington is "flatness," and we will go into this more deeply in chapter 10.

This leads us back to Ernst Mach, who suggested that the weakness of gravity was due to the universe's enormous size. Surely Mach could not even have contemplated such measurements in 1887, but his aspiration to formulate all the laws of dynamics by means of relative movements (with respect to all other masses) turned out to be visionary. Einstein was guided to general relativity by that very idea. Although Einstein gave Mach due credit,[10] he didn't ultimately incorporate Mach's principle in his theory. Mach's central idea, that inertia is related to distant objects in the universe, does not appear in general relativity.

One possibility to realize Mach's principle is to make up a formula where the gravitational constant is related to the mass and size of the universe. Because the universe is so huge, G would be very tiny. This radical idea was advocated by the British-Egyptian cosmologist Dennis Sciama in the 1950s,[11] but later only "outlaws" such as Julian Barbour pursued it.[12] During the inauguration of the Dennis Sciama building at the University of Portsmouth in England in June 2009, everybody talked about all the work that Sciama had done...except for Sciama's reflections on inertia.

It is idle to speculate whether Einstein would have warmed up to the idea. In any case, the size of the universe could only be guessed at 15 years after the completion of general relativity. So today, Mach's principle has a miserable reputation and is sometimes even dismissed as numerological hokum. To the arrogant type of theorists who consider the question of the origin of inertia as obsolete chatter from Old Europe, I recommend that they look up "Inertia and fathers," by Richard Feynman, on YouTube.[13] In this video, Feynman wonders about one big question. Where does inertia actually come from? A truly fundamental problem.

REAL EXCITEMENT OR A DELUDING PERCEPTION?

In physics, "*dot G*" describes how G changes with time. In the German language (and Mark Twain had already complained about its word rearrangements), "*dot G*" is translated into a synonym of "*G-spot.*" For this reason, nonphysicists might think that researchers are taking great interest in female anatomy, when the discourse is actually about the temporal change of G—which is pretty sexy to physicists, if not to anyone else. Puritans among the theoreticians are rather straight-laced about the idea that "big G" changes over time, while others are convinced that it does, just because of the pleasure they get from looking for it.

Eddington's hypothesis about the size and mass of the universe was picked up by Paul Dirac in 1938. (He put an even deeper reasoning behind it, but we will get to it later.) Dirac had been a celebrity in physics since the age of 26, when he applied Einstein's Special Theory of Relativity to the basic equation of quantum mechanics found by Erwin Schrödinger. Dirac suggested, in the spirit of Ernst Mach, that the expansion of the universe could affect the gravitational constant, because a larger horizon (the visible part of the universe) contains more mass.

By the way, Dirac developed the proposal of "*dot G*" during his honeymoon, upon which the cosmologist George Gamow dryly commented, "This happens when people marry." But at a mature age, in 1968, Dirac expanded on his earlier ideas about the gravitational constant changing over time, putting another log on the fire:

> Theoretical workers have been busy constructing various models for the universe based on any assumptions that they fancy. These models are probably all wrong. It is usually assumed that the laws of nature have always been the same as they are now. There is no justification for this. The laws may be changing, and in particular, quantities which are considered to be constants of nature may be varying with cosmological time. Such variations would completely upset the model makers.[14]

THE MOON, THE WAVES, AND THE BABYLONIAN ECLIPSE

Dirac's hypothesis about the changing nature of Nature continues to fascinate many physicists, and considerable efforts have been undertaken to prove that G indeed varies—up to now, unsuccessfully.[15] The first useful data came from the Viking space probe, which landed on Mars in the 1970s. Using radio signals, the distance to our neighboring planet could be determined accurately to within a few meters. A change of G as predicted by Dirac should have shown up there.

To the Moon, one can measure the distance even more precisely, thanks to the mirrors left behind by the Apollo missions and Lunar Laser Ranging. However, a "dirt effect" spoils the analysis. The Moon causes tides on Earth, slowing down its rotation due to the friction of tidal waves hitting the coasts. The tiny slowdown of the Earth is barely measurable, but there is a trick that uses the law of the conservation of angular momentum, which makes figure skaters spin faster by pulling in their arms. A slower-spinning Earth releases the Moon a little, increasing the distance between the two by just a few inches. This is measurable and allows us to calculate how much Earth's rotation was slowed down by tidal friction.

But do we have another independent measurement of Earth's rotation over the past millennia? Yes, in fact there is a beautiful collaboration of physicists and historians! Solar eclipses have been awe-inspiring events in all highly developed cultures, and they were scrupulously recorded by the Babylonians, the Egyptians, and the Greeks. In China, a drunken astronomer is said to have been beheaded for having missed announcing this sinister occultation to his emperor.

By studying the ancient chronicles, astronomers can date about 200 solar eclipses.[16] For instance, one in Babylon (present-day Iraq) proves that Earth's rotation indeed became slower. Otherwise that eclipse, which is easily calculated, would have gone through the Spanish Mediterranean, 2,500 miles off of Babylon. But still, a problem remains. Even if Earth's rotation decelerated by the amount we are told by the moon's distance, the eclipse should have occulted a stripe 1,000 miles west from Babylon. This indicates that there might be a change of the gravitational constant (remember, this also affects the distance to the moon), or even more puzzling, an anomaly in the flow of time (which would affect the rotation).

CROSSED WITH *G*

It is claimed sometimes that life could not have existed at all in our universe if the fundamental constants did not have the exact values that they do. We will later try to extract the interesting part of this statement from the ramblings, but one thing must be said beforehand. We could all live happily if the value of *G* alleviated all weights by, say, 20 percent, except for a few orthopedists...and the majority of theoretical physicists.

Dirac's provocative words in 1968 apply today more than ever. A change of the gravitational constant is a nightmare for most, since it would pull the rug out from under the standard models of cosmology and particle physics. For this reason, I remain somewhat suspicious when researchers claim to have seen the constancy of *G* in a primordial hydrogen structure or similar tea leaves.[17] We can't be sure. But precise clocks will eventually bring it to light.

Looking at the history of science, our belief that we have found the ultimate gravitational law seems a little naïve. How much time had to pass before the idea of local gravity g (9.81 m/s^2) took hold? And how many battles did Galileo and others have to fight before establishing today's gravitational law based on G? To recognize the relevance of the gravitational constant G for the entire solar system has been a tremendous achievement of astronomy.

Many scientists may have made fun of a certain geographically challenged American president, who praised a planned mission to Mars as the "conquest of outer space." However, the practitioners of gravitational physics are themselves rather starry-eyed by blindly trusting G. The solar system has provided evidence for the gravity theories of Newton and Einstein, but it is a mere islet when we envision the dimensions of the galaxy or the universe.

Chapter 6

THE RIDDLE OF SMALL ACCELERATIONS

ARE GALAXIES REALLY JUST BIG PLANETARY SYSTEMS?

In the 400 years since Galileo first pointed his telescope at the sky, we have been able to study planetary orbits in our solar system in considerable detail. Mercury holds the record with nearly 2,000 revolutions around the Sun during the past four centuries, while Pluto, now downgraded to a dwarf planet, has yet to complete a single orbit around the Sun since it was discovered in 1930.

All these motions are dictated by the Sun's gravity. Galileo also demonstrated that Jupiter was the center of its own gravitational system. He was the first to observe Jupiter's moons and to determine their time of revolution. This time, called the "orbital period," is a precise measure of the mass of the orbited body. The bigger the mass, the shorter the orbital period, and the faster the moon.

Although somehow obvious, it is nonetheless a remarkable idea to use this planetary clockwork for mass estimates on the scale of galaxies. Of course, we can't sit and wait for the millions of years the stars need to make a single revolution around the galaxy center, but fortunately, the Doppler shifted light emitted by distant stars quite accurately reveals their orbital velocity.

Apart from some widely ignored radio astronomers, the first to make use of the Doppler shift to determine the mass of a galaxy was Seth Shostak, when he observed[1] the disc-shaped, rotating spiral galaxy NGC 2403. (By the way, Shostak dedicated his doctoral thesis "to NGC 2403 and its inhabitants, to whom copies can be furnished at cost." Later, he even became a key figure in the Search of Extraterrestrial Life [SETI] radio astronomy program.) Shostak's research in the early 1970s revealed the groundbreaking problem we are talking about in this chapter: the large orbital velocities of the stars in the galaxy he was studying indicated the existence of much more mass—what would later be called "dark matter"—than was expected from star counting!

Meanwhile, the Princeton cosmologist Jim Peebles, along with colleague Jeremiah Ostriker, had been struggling with their model of the Milky Way. They had assumed that it was a disclike mass distribution of the visible stars. According to their computer simulations, the galaxy disc should have quickly destabilized and broken up, with the stars flying off into space. Their model could only be fixed by hypothesizing a large, spherical mass distribution around the galaxy called a dark matter "halo."

The astronomer Vera Rubin from the Carnegie Institute in Washington picked up on Seth Shostak's idea that something was wrong with the mass of galaxies—the stars were moving too fast. She concluded that perhaps astronomers were underestimating the weight of galaxies by a factor of ten. The suggestion generated outright hostility. Much later, Rubin was honored for her work with the Bruce Medal. The Dutch astronomer Robert Sanders commented in his excellent book *The Dark Matter Problem:* "Sometimes one is too much ahead of the pack."

NO WAY FOR NOT SEEING: GALAXIES LACK MASS

If we look at a spiral galaxy edge-on, we see a relative blueshift of the side that is approaching, while on the other side, the rotational velocity points away from us and the light is shifted to the red. If you then measure the outermost points of the spiral arms, you will right away obtain the mass of the galaxy. This is higher than the mass estimate based on the luminosity of the galaxy, assuming it consists of stars like the Sun. However, such a census that assumes every star is like the Sun isn't very precise.*

Rubin's critics tried to talk their way out of the dark matter problem (back then called "missing mass") by inventing creative excuses for this anomaly.

---

* A star, by doubling its mass, becomes about ten times as luminous. This makes mass estimates difficult, because the size distribution of stars, called initial mass function, is hard to guess.

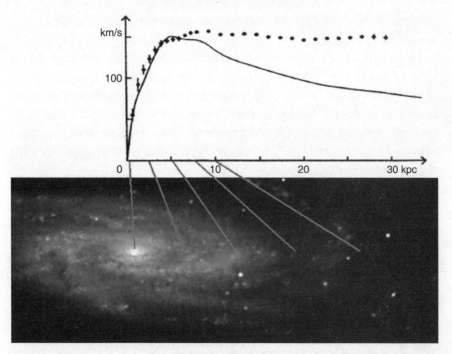

Fig. 7. Rotation curve of a typical spiral galaxy (NGC 3198), where the rotational velocity is plotted versus the distance to the galaxy center. The solid line is expected from Kepler's law of planetary motion, contradicting a typical measurement (dotted line). Therefore dark matter is hypothesized in the region, where the curves don't fit together. The connecting straight lines indicate that rotation curves can be recorded far away from the visible range of 10 kiloparsec (approximately 30,000 light years).

Eventually, the radio astronomers came to the rescue and supported Rubin with their findings. They could measure the speed of gas clouds beyond the visible spiral arms, and this was a really unpleasant surprise waiting for the optical astronomers. Clouds weigh almost nothing, and hence the velocity of such gas clouds had to decline with increasing distance from the center (fig. 7.).

Instead, the velocities remained practically the same once they had reached their maximum value. This is often called a "flat rotation curve."

Flat rotation curves suggest that the outer parts of galaxies contain mass that is not visible. The anomalous flat rotation curves have been observed in almost 1,000 spiral galaxies.[2] They are considered to be the best evidence for dark matter.

The shape of the rotation curve even tells us about the mass distribution. The density of mass declines the farther it is away from the galactic center; however, every spherical shell around the core of the galaxy seems to contain the same amount of mass, no matter what the distance is. This is also very puzzling.

In many galaxies, dark matter outplays its "normal" counterpart by a ratio of ten to one, sometimes more. This is strange enough, and recent observations reach farther and farther beyond the galaxy's visible part.[3] But a decrease of the rotation curve—implying the ending of the dark matter halo—has not yet been found anywhere. Thus, we seem to be in a situation where the quantity of dark matter in the universe keeps growing with the quality of our telescopes!

The anomalous rotation curves are, however, just one among many observations where the determination of mass runs into contradictions. Once the problem of missing mass had finally sunk in, a sort of counter movement started. Robert Sanders commented ironically, "With respect to the dark matter problem there existed a group dynamic that refused to accept the obvious evidence for a significant anomaly in galaxies and then, in a total reversal, saw dark matter everywhere."[4]

THE INTERGALACTIC THERMOMETER

Evidence of a very different kind comes from satellite telescopes that show us X-rays originating in the distant universe. Looking at galaxy clusters, we see diffuse radiation around their centers, which indicates extremely hot hydrogen gas. According to Planck's law, the small wavelengths of X-rays can only arise from high temperatures. But temperature is just another name for the average kinetic energy of particles, and it is strange that such energetic particles don't leave the galaxy cluster. Either they are just passing by coincidentally, which seems absurd, or there must be more mass involved than we expected from the cluster's luminosity that is keeping the hot hydrogen there—a clear trace of dark matter. By the way, this is Fritz Zwicky's old discovery in a miniature form. Instead of measuring particle velocities, he had measured the single galaxy's motion, but the riddle in both cases was why they kept to the cluster.

In a galaxy, stars fly about with varying velocities, and the so-called dispersion, a statistical figure for their speeds, states to what degree stars seek to leave the gravitational field of their mother galaxy. This velocity dispersion is very high for elliptical galaxies, and again—you probably guessed it—there isn't enough visible mass. This method can be used for globular clusters, actually miniature galaxies, as well. These beautiful accumulations of about 1 million stars each inhabit the suburbs of our Milky Way. Once again, a researcher from the European Southern

Observatory found unusually high velocity dispersions.[5] Can this problem be fixed the by invoking dark matter? No, it doesn't work. The much smaller scale of star clusters implies that there may be only negligible traces of dark matter there, which is quite wondrous. And objects just a little bigger, dwarf galaxies, recently have posed serious contradictions to the common model.[6]

MASSIVE COMPACT HALO OBJECTS:
THE DYING-OUT MACHOS

Although all the observations leading to the "missing mass" idea are well confirmed, there are still reasonable explanations that have warranted consideration. Small stars are known to shine much more weakly than larger ones, and they cannot be spotted across the Milky Way, not even with the best telescopes. Could most of the missing mass be contained in these dwarf stars after all? Unfortunately, this is too simplistic a view, because dark matter cannot be solely located in the galaxy's disc. If this were the case—and our galaxy is arguably a typical specimen—then those approximately 200 globular clusters orbiting the Milky Way outside the disc plane would have different positions and motions.[7]

This and computer simulations have convinced astronomers that dark matter is located in a spherical "halo" that surrounds the galaxy. The natural guess was then that this halo was populated by "brown dwarfs," which are unremarkable celestial objects too small to ignite the fusion of hydrogen to helium, which would have birthed them into the luminescent existence of a star.

But astronomers had an idea for testing the hypothesis of so many brown dwarfs flying about. Although invisible themselves, the dwarfs can temporarily raise the brightness of a background star by slightly focusing its light. This is a form of gravitational light deflection called "microlensing." Immediately, an intense and systematic search began, but there were too few dwarfs found to account for the elusive missing matter.

One might think here about black holes, too, as possible small but massive objects, but astronomers don't consider this seriously. Black holes are in actuality not invisible, since their surroundings would emit X-rays. Thus, the assumption of Massive Compact Halo Objects, or "Machos" for short, has been on retreat. So far, all attempts to invoke conventional matter have failed to explain the phenomenology of its dark counterpart. Conventional matter does not translate into dark matter; all feasible possibilities seem to have been ruled out.

WEAKLY INTERACTING MASSIVE PARTICLES:
THE WIMPS ARE COMING

Whenever possible, substitute constructions out of known entities for inferences to unknown entities.

—Bertrand Russell

Traditional explanations for dark matter tended to garner more sympathy among practical astronomers. Theorists, instead, seem to prefer wholly
new types of particles. There is no shortage of ideas about how hypothetical particles could nicely fit dark matter's behavior. Unfortunately, none of
them has been found in any lab. Since the properties of these new particles
are utterly unknown, it's quite easy to argue that they don't reside in the
galaxy's disc like normal matter does, but rather hang out in the halo. The
halos surrounding galaxies have become huge screens upon which theorists are projecting their hopes for new particles.

A favored theory for new particles is "supersymmetry," affectionately
known as SUSY. It was particularly creative in inventing hypothetical
particles, with neutralinos, photinos, axinos, and W-inos* being the top
candidates. Considering the hasty proliferation of particles showing up in
the literature, "weakly interacting" could just as easily be labeled "weekly
interacting."

To date, all these hopes have vanished into thin air. Robert Sanders hits
the nail on the head: "The real problem is that dark matter is not falsifiable.
The ingenuity and imagination of theoretical physicists can always accommodate any astronomical non-detection by inventing new possible dark
matter candidates."[8]

GRANDIOSE PREDICTIONS, NOT TO BE FORGOTTEN

It's fine to have grand theoretical ideas, but it does seem like a good idea
to connect the phenomenon of dark matter with real observations. Take,
for instance, the Bullet Cluster, an unusual object astronomers have been
studying since 2006. It is actually made up of two colliding galaxy clusters. This seems to be a dramatic event, but the single galaxies passed by
each other much as ships of a fleet pass in the ocean. The huge, hot gas
clouds within the cluster, however, were stripped off during that encounter. While the clouds could be identified by X-ray emission, the analysis

---

* Maybe not coincidentally, there is a phonetic similarity to vino, the Italian word for wine. Italian
diminutives however seem to be an inexhaustible source of hope for particle physicists.

of background light (gravitational lensing) seemed to indicate that dark matter, as the galaxies, had passed without colliding.

This was, undoubtedly, an interesting observation, hyped as a "direct detection" of the invisible and unknown dark matter. The interpretation, however, does depend on several assumptions, such as the tendency of the gas to emit X-rays and the amount of matter in the gas clouds. It is far from being unambiguous yet.[9] And the analogous case of the "Train Wreck" galaxy cluster (the name may come from the headaches it created for dark matter advocates) shows merely one thing—that we do not yet understand these collisions.

On the other hand, many predictions of dark matter have fizzled out. For example, people were assuming those famous WIMPs (weakly interacting particles) sometimes annihilate, producing two gamma quanta. This gamma radiation should have been detected by powerful X-ray telescopes such as FERMI. According to calculations of a group led by Carlos Frenk from Durham University, the radiation should have revealed the dark halo of the Milky Way. Perhaps to avoid accusations of false modesty, Frenk discreetly noted that such a discovery is surely worth a Nobel Prize. Nice that we know the value of the fur, but we are still waiting for the bear to be bagged.

Should FERMI continue to see nothing, I am curious if this will be interpreted as evidence *against* dark matter, or if we will once again hear the excuse that the particles interact even more weakly than theorized. "There is still the possibility, that nature refuses to cooperate," said one theorist.[10] We shall hope that people do not forcibly crowbar the elusive dark matter into the interpretation of future observations.

MESSING WITH THE GRAVITATIONAL LAW— HOW NOT TO DO IT

To a man with a hammer, everything looks like a nail.

—*Mark Twain*

For 80 years now, people have been chasing dark matter without getting any the wiser about what it might be. Could this be the story of phlogiston repeating itself? Phlogiston was believed by early chemists in the eighteenth century to be the volatile substance in materials that burns and is given off in the flame. They never found this fire-substance, although the experimental search for it did play a key role in the development of modern chemistry.

It is perfectly sensible to question whether there is something mistaken about our understanding of gravity rather than continuing to search for a modern-day equivalent of phlogiston. To that end, theoretical physicists have put forward several modifications of Newton's law of gravity.

An obvious guess was whether the law of gravitation might change with distance. Messing about with Newton's law in such a way, however, just illustrates how thoughtlessly physics can be practiced at times. Such a modification is not only hideous, it also falls flat on its face when one compares it to the data. Dark matter—let's call it "the anomaly" so as not to limit it to being matter—does not occur at a specific radius of the galaxies, but in a halo at roughly the position of the outermost stars, whatever distance from the center they might have.

Galaxies vary greatly in size, differing by a factor of 100. Put to the acid test, these modifications of gravity have the embarrassing implication that stars in certain galaxies ought to be 100 times less bright than they normally are. "Rather unsatisfying for a model constructed to remove the need for dark matter," as the astronomer Anthony Aguirre pointedly noted in a research article.[11] It was a kind of tutoring for theorists who meddle with the laws of Nature without knowing the observations.

Playing with the law of gravitation has become a mass sport, with arbitrary inventions almost worse than those of particle physicists. In taking the maximum liberty for new formulas, beautiful complications can be produced. Look, for example, at the so-called $f(R)$-gravity: it modifies the curvature of space R in the general theory of relativity with *any* function (f). Well, why not? Such an idea was proposed at a cosmology conference in Munich in 2008 as a means of fixing a contradictory observation. The cosmologist Ruth Durrer from Geneva raised her hand and commented dryly, "This modification of gravity is indeed interesting, but it would be even better, if it agreed with the solar system orbits!" The lightweight tiny theory had literally lost ground.

Although Einstein's rebuilding of Newton's laws was very successful, it's hard to see how it should cope with all the anomalies that have piled up in the last decades, in particular the puzzling dark matter phenomenology. General relativity becomes important in strong fields, but Newton's and Einstein's theories are identical for small accelerations. If we confirm an anomaly *there*, it would be especially unsettling because it would imply that there is a flaw in both theories!

MOND: THE CONSULTANT FOR ASTRONOMERS

Sometimes it helps to stand back and take a detached look at a problem. At its heart, the idea of changing the law of gravity at a certain radius is utterly naïve. What matters in orbital motion (just as for race cars speeding around curves on the racetrack) is centripetal acceleration, and it makes much more sense to try a modification of gravity depending on how strong a body is accelerated.

However, it wasn't astronomers who first pushed this idea, but Mordehai Milgrom, a solid state physicist at the Weizmann Institute in Israel. When he came up with his proposal MOND (a coinage for MOdified Newtonian Dynamics), to change the law of gravity at small accelerations, people were skeptical, to put it mildly. Several science journals rejected his manuscript as a fallacious idea, but he was lucky that he eventually got it published by the reputable *Astrophysical Journal* in 1982. His formula described the flat rotation curves of spiral galaxies surprisingly well.[12]

Milgrom assumed that, for small values, the common gravitational acceleration g would be substituted by a mix* of g and a_0, with a_0 postulated to be a "fundamental" acceleration. Based on this, as virtually every expert on rotation curves admits, MOND is an excellent phenomenological description.[13] Though the parameter a_0 was fitted to match the data, this one single number describes the behavior of many hundreds of spiral galaxies that vary in mass by a factor of 1000! Even if this ad hoc mending by a_0 may be a questionable method, the baffling numeric value has attracted interest. If you divide the speed of light c by the age of the universe T_u, you get a value for acceleration in the same order of magnitude[14] as a_0.**

Before sounding like we are descending into numerology, I shall clarify. MOND is a terribly poor "theory" and hardly deserves the name. It abandons everything sacred to gravitational physics, from Newton's law of reciprocal actions to Einstein's principle of equivalence. Although recently a patch version of MOND was released that seems to be less damaging to theoretical physics, most experts still get sick at the thought of this. Therefore, many astronomers are gleeful to see the predictions of MOND fizzle out, because it can neither explain the missing mass in elliptical galaxies, nor the dark matter in galaxy clusters—not to mention its failure to explain the data of the early universe.

IF B IS WRONG, A MUST BE CORRECT

If an idea is not dangerous, it is not worth calling it an idea.

—*Oscar Wilde*

Many scientists, especially those who consider themselves openminded, mention MOND in their talks as an alternative to the standard model of

---

* Technically, the geometric mean.

** Similar to the anomalous acceleration of the Pioneer spacecraft, which however seems to have a satisfactory conventional explanation, though some questions remain.

cosmology. Sadly, sadly, though, where there is so much evidence contradicting MOND, scientists return to the standard model and continue to use
its blinders. I have never fully understood this kind of logic. To say it again,
MOND is wrong, and I would be the last one to shed a tear over it. But it
undeniably has one merit. MOND revealed that anomalies in spiral galaxies appear at small accelerations. It is an astonishing and irritating insight
that Newton's law had not been tested for that small acceleration range, and
we owe it to MOND to give credit where credit is due. And even if MOND
founders when it comes to galaxy clusters, there is one thing in common at
all appearances of dark matter: its motion is observed at tiny accelerations,
mostly just below the magical value of c/T_u, derived from the speed of light
and the age of the universe. It is precisely this acceleration that was found
in (globular) star clusters,[15] although dark matter technically should not
exist there.

Whether you want to believe in a relationship between "the anomaly"
and the universe's age or not, one thing is sure. Either 1,000 spiral galaxies
agreed on fooling us with their concerted behavior, or we don't understand
the fundamentals of gravitation yet. The data clearly suggests the latter
option. But, as Lee Smolin has noted, there is a psychological barrier at play
in the physics community: "The dark-matter hypothesis is preferred mostly
because the only other possibility—that we are wrong about Newton's laws,
and by extension general relativity—is too scary to contemplate."[16]

IGNORING THE WARNING SIGNAL—THE CARAVAN CONTINUES ON

I can't help but tell you how some leading scientists react to MOND's findings. The Max Planck Institute near Munich is one of the most prestigious
research facilities in astrophysics. Thus, I was looking forward to attending
an open house there, where scientists from all over the world would answer
any question. As a teacher and, in this respect, a representative of poor,
educationally distressed schools, I am always welcomed by experts who
generously pass down their elite research.

Since I happened to sit at a table next to Simon White, the institute's head,
I asked what his opinion of MOND was. "I have no *opinion*," he replied
thoughtfully. "As a scientist, I only look at the data." Yet he looked like I
had just proposed to read his horoscope, and I was thus quick to assert that
MOND was, of course, refuted and theoretically horrifying. But I asked
whether we should nevertheless wonder about all spiral galaxies carrying
the particular acceleration c/T_u related to the age of the universe. "I regard

it as pure coincidence," he said, "and besides, the distant galaxies should show us then a completely different age of the universe."

I tried to bring to mind for a second where I had heard that argument before. I eventually remembered that, several months earlier, I had told my astronomy course about the dark matter riddle. One of my best students—he had twice won the "contest for young scientists" with his creative aircraft constructions—soon came up with the same argument against MOND. Indeed great distances, taking into account light travel time, are a glimpse of the past, and one would have to consider the galaxy's erstwhile age when calculating the acceleration c/T_u.

On closer inspection this doesn't, of course, argue against MOND by any means. Rotation curves, a subtle and difficult measurement, can only be attained from relatively nearby galaxies—otherwise we would already have inspected more than 1,000 of the universe's assortment of 100 billion galaxies. The analyzed galaxies are therefore quite representative, but altogether much too close for the light travel time to play a role. My student back then understood this right away, and hence I did wonder a little about White's argument. While I was replaying the story in my head, he had become involved in another conversation and apparently felt no further need to discuss the intricacies of MOND.

BETTER TO FADE OUT THAN GIVE UP

Some months later, I began to think that my impartial question may have annoyed White a little. His research is among the most cited by astrophysicists. Hundreds of articles are devoted to a "Navarro-Frenk-White" profile of dark matter in galaxies, and extensive simulations by supercomputers are based upon this. My naïve question to him about MOND ultimately implied, "Might it be that you have devoted the last 25 years of your research trying to catch a phantom?"

Let us be honest: for anyone in White's situation, it would have been hard to answer such a question without any emotion. Being mistaken is part of the professional risk of being a scientist, but it can also be painful enough that objective indications become too easily disregarded. The mind's repression probably makes sense for the individual, but is nonetheless a debilitating mechanism for science as a whole. Maybe the biggest methodological problem of physics lies herein—it is the fact that scientists are humans.

The small accelerations are, in contrast, *the* content problem of gravitational physics. Starting with the measurements of the gravitational constant and ending with the expansion of the universe, there is no single observation that hasn't run into difficulties below the tantalizing value c/T_u. Instead of fantasizing about length scales of 10^{-35} meters, physics has to understand accelerations at the scale of 10^{-10} m/s^2. In contrast to Planck's length, something can be actually seen here. However, one has to be willing to look.

Chapter 7

LOST IN THE DARK

DARK MATTER AND DARK ENERGY: INVISIBLE OR ALL IN YOUR MIND?

The mysterious phenomena occurring at low accelerations are usually associated with "cold" dark matter (CDM), which means its particles are supposed to move slowly. To incorporate the observations regarding dark energy, the standard model of cosmology has now been upgraded to a "Lambda CDM" model. However, there are still a lot of observations challenging it.

Take, for instance, the story of "low surface brightness" (LSB) galaxies. Astronomers have long thought that galaxies would clearly stand out from the background space due to their luminosity. This preconception was proved wrong by Gregory Bothun of the University of Oregon, who used a sophisticated technique. He analyzed the radiation with a wavelength of 21 cm,[1] which is the fingerprint of hydrogen, and discovered a spiral galaxy of incredible size. It is twenty times larger than the Milky Way but invisible to conventional telescopes!

It now appears that astronomers have underestimated the amount of mass in the universe due to the hard-to-find LSB galaxies. It is possible that a great many galaxies exist at places in space where we don't see anything…yet. This one small example illustrates the point that we should expect the unexpected in astrophysics. In the end, LSB galaxies could also emit some of the radiation we currently believe to originate from the cosmic microwave background. The data of the WMAP and Planck spacecraft would then turn out to be far less accurate than we think.

STILL TO BE UNDERSTOOD: THE DYNAMICS OF GALAXIES

A dwarf galaxy (which is a funny name for a collection of a billion stars) sometimes has a low surface brightness, and this gives galaxy experts serious headaches. The rotation velocity of their gas clouds indicates the presence of dark matter, but a hundred times more abundant than in usual galaxies. This is hard to explain by means of the behavior of any fundamental particle.

In general, these rotation curves (see fig. 7) show a highly developed regularity. With large spirals, the velocity maximum for stars is reached at the visible edge of the galaxy, and then slightly drops. In dwarfs, however, the rotation speed of gas clouds continues to rise even far beyond the visible edge. Dark matter seems to prefer the outer regions of dwarf galaxies.

Enthusiasts of darkness, such as Dan Hooper in his book *Dark Cosmos*, don't worry too much about this weird fact. But those who scrutinize the rotating spiral galaxies, like Paolo Salucci of the University of Genoa, are increasingly annoyed, when theorists cook up an explanation for the allegedly flat rotation curves by means of a new dark matter candidate particle. If dwarf galaxies didn't exist, the problem of dark matter would be "*molto, molto minore*,"* as he wrote to me in an e-mail.

Others, such as Jerry Sellwood[2] of Rutgers University and Wyn Evans[3] of the Institute of Astronomy in Cambridge, have described how the conservation of angular momentum and clearly visible bars in some spirals contradict a dark matter explanation regardless of whatever particle is proposed.

Mike Disney of Cardiff University recently gave an insightful overview of observations in galaxies.[4] Evidently they are a lot simpler than expected. Galaxies differ in mass, size, angular momentum, dark matter ratio, chemical composition, and age, but all of these quantities are interrelated in an inexplicable way. It is as if the theory predicts a colorful circus assemblage milling around, and instead we are observing a uniformed military elite unit in lockstep. There is no conventional explanation for this. Disney has, by the way, a rather harsh opinion of the gravitational law outside of the solar system: "It simply never works."[5]

DO PTOLEMY'S PERFECT CIRCLES ALSO EXIST IN GALAXIES?

Maybe we have to abandon our perception of galaxies as amplified solar systems. When you look at pictures of gorgeous spirals such as M51 on the

---

* "Much, much minor."

Internet, you might see the problem for yourself. If everything is orbiting along circular paths, but the speeds vary for different radii, how is the spiral shape upheld? Surely, we can't be looking at stiffly rotating discs in the style of a vinyl Beatles album.

This problem is considered solved by the theory of so-called density waves. According to this idea, stars themselves do not move with the spiral arms of the galaxy. Rather, gravitational blast waves sweep through the hydrogen gas in the spiral arms, giving "birth" to bright, blue stars. These only light up a couple million years and then go out. So, we are, metaphorically, seeing white crests of waves in an ocean as structures, not the water itself. It is somewhat odd that galaxies are still not "sucked dry" of available hydrogen gas after many circulations of the gravitational blast waves, since they seem to be producing stars with almost the same rates ever since their emergence.

However, if one agrees with Disney's doubts concerning the gravitational law, then one also has to challenge the doctrine that all motions take place on steady orbits with unchanging radii, since we can't actually *see* this. Astronomical observations are snapshots that are blind to motion passing by. We can only see velocities in the line of sight that are revealed by the Doppler shift. One of the first "real" measurements of the velocity of objects that move along with us (as opposed to away from or toward us) was published some time back, made possible by images of the Magellanic clouds taken with a ten-year gap in between.[6] The result? The galaxies that are satellites of our Milky Way are moving at such high speeds they seem to be flying past us. Is that evidence that they need to be kept by an extra pull of even more dark matter? Or are we, yet again, not fully understanding it? It is worth noting that the positions of these satellite galaxies contradict the idea of a dark matter halo. They lie in roughly one plane, while the halo model predicts a random distribution. A quite unbelievable accident.[7]

Researchers do have their preferred models, and you may say that it's no surprise that data is interpreted differently. But even if some experts on galaxy dynamics may be on bad terms with each other, there is no one who really embraces the dark matter model—it just fits the data badly. And with respect to galaxy formation, the astronomer James E. Gunn of Princeton said: "I'm sure it's wrong. It's too simple. Nature is never that cooperative. But it's one of the only schemes that allow you to do calculations."[8] Argyro Tasitiomi of the University of Chicago mockingly summarized the three kinds of expert opinion[9]:

1. "The dark matter model is too compelling to be wrong."
2. "There is something wrong with the dark matter model."
3. "The model of dark matter is wrong."

It is certainly annoying when off-reality superstring luminaries such as Leonard Susskind announce in a YouTube interview that dark matter is "not mysterious at all," but rather "just particles."[10] This rather suggests that Susskind is "just clueless" about galaxy dynamics.

WHAT IS DARK MATTER? GOOD QUESTION. NEXT QUESTION?

Science is built up of facts, as a house is with stones. But a collection of facts is no more a science than a heap of stones is a house.

—*Henri Poincaré, French mathematician*

Don't be mistaken in thinking that physicists devote themselves to agonizing over dark matter. You will find more creative ideas by Googling "dark *energy* theory" than "dark *matter* theory." Scientists are simply already used to dark matter. It is more accepted because the phenomenon of dark matter is based upon a series of measurements, while the notion of dark energy rests in essence upon observations of the too faint and distant supernovae.*

Dark energy is the name given to the mysterious energy dominating the universe and causing it to expand at an ever-increasing rate. A famous *Nature* issue in 1998, with a cartoon of a smiling Einstein smoking a pipe, celebrated dark energy as the revival of Einstein's cosmological constant "Lambda"—surely, a grand promotion of something better called "a lack of understanding."

It's not clear whether Einstein would be happy about being portrayed as a crown witness for dark energy. In 1917, he had launched the mathematical complication Lambda with a heavy heart in order to save his physically simple model of the universe, precisely because he was convinced of Nature's simplicity. Today's Lambda is merely one of the many ingredients of the standard model, a roadman's pebble rather than a philosopher's stone.

But even if one sees in the cosmological constant a late reverence for Einstein, his conceptual parsimony at the time has become a brake for the greediness of modern theorists who want to inflate their models with new numbers. Instead of one single value such as Lambda, scientists now favor "quintessence,"** a field everywhere in space. This means theorists can adjust the values for different places in space, depending on what

---

* Apart from the sociological now-everybody-sees-effect.

** The term is a throwback to medieval times when people believed the world was made up of four elements—earth, air, water, and fire—and a fifth all-pervasive substance throughout the heavens called "quintessence."

observation they want the theory to fit. But you don't need to be a specialist to realize how much arbitrariness is introduced into a theory with such a truckload of freely selectable parameters.

Keep in mind that Einstein had, after years of pondering, come up with *one* new number. And Max Planck's new constant h, the first step toward quantum mechanics, had him tumbling into self-doubt. Planck probably turned over in his grave when, in 2005, the research prize named after him was awarded for—wait for it—quintessence!

There is a species of physicists claiming that fields like quintessence are more "natural" than a single number such as Lambda, since such abstract fields are part and parcel of elementary particle physics, too. Transporting these bad habits into cosmology is as silly as claiming that horribly distorting subsidies are generally good, because they work so well (yeah, right) in agricultural economics. This fashionable fad for fields says more about the nature of particle physicists than about the physics and particles of Nature.

DARK ENERGY, QUINTESSENCE, OR FAINT SUPERNOVAE?

Theories crumble, but good observations never fade.

—*Harlow Shapley, American astronomer*

Instead of climbing the ladder of theoretical hypotheses, science would benefit by coming closer to the actual observations of raw data.

The supernovae detected by the teams who got the 2011 Nobel Prize *are* too faint with respect to their redshift, and now the true supernova luminosities are known much better due to the award-winning new method. However, the brightness of these star explosions still varies greatly. To mitigate this nuisance, in several publications on supernovae, you will find "binned data," which amounts to averaged data points packaged to make a nicer fit to a line on a graph. The procedure reminds us just a bit of the practice of securitization by repackaging mortgage loans, a questionable ironing out of physics by statistics.

I'm not suggesting that the supernova data is not significant as a whole; it clearly contradicts the old model of an expansion of the universe that slows down. But an accelerated expansion of the universe raises new questions. It seems as if the supernova measurements were also compatible with the model of an "empty" universe, which would have to mean that gravitation wouldn't act at all, a tantalizing possibility that cannot be integrated into any theory.

But maybe such dangerous ideas merit interest. The dark energy model, as pacifying as it is, produced the coincidence problem, as noted in chapter 3.

By some unexplained coincidence, we apparently live in a really extraordinary era of our universe in which dark energy gains control over the contracting forces. This should be matter for reflection.

Moreover, cosmological expansion takes place on an extremely long time scale (the age of the universe being T_μ), while the most distant objects we observe seem to recede with a velocity close to c,* the speed of light. Therefore, the universe's expansion is yet again formally linked with the notorious acceleration of c/T_u, which has already presented us with quite a few puzzles.

At this point, it is useful to clarify that, apart from being referred to as "dark," dark matter that shows up at c/T_u and dark energy have nothing to do with each other! You may argue here that contemplating c/T_u does not bring you closer to a conclusive theory, but before starting, it is important to realize that we do not fully understand gravitation on a cosmological scale.

The time span granted to our civilization so far provides us with little more than a snapshot of the universe. Thus, a great deal of fantasy and even more optimism is needed to tell the story of the universe's origin, as many authors of popular science books boldly do. The history of cosmology has every now and then confronted us with a radical change in our world view. It would be utterly naïve to think that we finally got it right some 15 years ago.

THE MYSTERY OF THE TWO-DIMENSIONAL UNIVERSE

This headline is not a counterthesis to the high-dimensional fantasies of string theory, nor is it a way of ridiculing a certain idea that wants to combine gravitation in two dimensions with quantum theory (because its adherents find it too difficult to do in three dimensions). No, it is a matter of real observations.

Andromeda is, at a distance of 2 million light years, the galaxy most visible in the Northern Hemisphere. It is a decent example of a spiral galaxy, which, together with the Milky Way and two dozen smaller galaxies, form the "local group"—comparable to an extended tribal family that lives in a couple of scattered huts. The next "tribe," the Virgo galaxy cluster, is 60 million light years away, but all this you can look up easily.

The really interesting question is, "What structure does the galaxy distribution in the universe have, if any?" This, too, provides a test of whether our models of gravitation are correct or not, given that gravity is the responsible force for structure formation.

---

* General relativity would require a more subtle discussion, which is however irrelevant for our purposes here.

Apart from groups and clusters, astronomers have also identified galaxy super clusters, concentrations of several thousands of galaxies. Although this sounds like an appropriate classification, Gerard de Vaucouleurs, a rival of Hubble's scientific grandson Alan Sandage, wrote thoughtfully as early as 1970, "Discovering super clusters, we have once again believed, that these are the greatest structures of the universe, as we once did after finding clusters, groups, or galaxies themselves."[11]

While previous research on cosmic structures relied merely on guess-work, in the 1980s, John Huchra and later Margaret Geller, started to sys-tematically measure the distances through redshifts and thus obtained the first three-dimensional picture of the distribution of galaxies. Their sensa-tional article "A Slice of the Universe"[12] revealed a fundamental property—the universe looks like a colossal sponge (see fig. 8).

Next to the huge empty space bubbles called *voids,* matter is concen-trated in two-dimensional structures of galaxy clusters, or rather, galaxy meadows. Ultimately, we do not understand this cosmological lather, and no one has predicted it either. Even Margret Geller, despite her significant findings, remains quite skeptical. "I feel often that we are missing some fundamental element in our attempts to understand the large-scale struc-ture of the universe."[13]

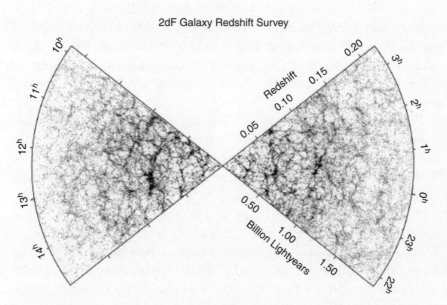

Fig. 8.   Modern reconstruction of the galaxy distribution in a disc-like surround-ing of the Milky Way, as created by Huchra and Geller in 1986. Clearly visible are the inhomogeneity and the vast empty spaces.

Furthermore, there is the following mystery: if one considers the circular "surfaces" of the disc-shaped galaxies and adds them all up, the result is an enormous value. It is—and therein lies the strangeness—in the same order of magnitude as the surface of a sphere with the radius of the visible horizon (which is another name for the universe's "surface"). This galactic surface estimate may well be vague, but the radius of the horizon is continually growing due to the Hubble expansion, while the dimensions of galaxies have remained constant for billions of years, at least according to the prevailing belief. Another unsettling coincidence of present time.

LOCAL HUBBLE LAW, FOR LUNCH

Evidently, *any* structure in the distribution of galaxies is a deviation from Hubble's law of uniform expansion of the universe, because the observed groups and clusters could have only developed through disturbances of the allegedly homogeneous original state. Gravitation, which enhances the clustering of random fluctuations, could be the obvious reason for this.

If we want to reliably measure the expansion rate (the Hubble constant), we need to do it over great distances in order to average out local irregularities. At small scales, we would expect random motions to predominate. However, when performing the measurements in this allegedly messy neighborhood of our local galaxy group, the precise value of the Hubble constant pops out, even though the Hubble expansion can hardly be at work![14]

It was very interesting for me to have once witnessed how such problems are discussed among cosmologists. David Wiltshire from New Zealand mentioned this riddle at the Leopoldina conference on dark energy in Munich in 2008. He promptly offered a solution, which, however, seemed somewhat ad hoc.

When the lunch break arrived, a group with Ruth Durrer and Roy Maartens* asked me for directions to the "English Garden," a huge park in the center of Munich, which is famous all over the world for its nude sunbathers. My colleagues were, however, more interested in whether there was a place to snack along the way.

But before I could show my suggested location, Ruth Durrer discovered a greengrocer in front of the university and bought lots of vegetables for the healthy lunch the Swiss are so famously fond of. Two Italians in the group looked at each other in despair, and an apparently growling stomach made one of them exclaim a "Io c' ho fame!"** before they accepted their fate.

---

*I was familiar with that name since my friend Sante told me in Rio: "*Uno con i coglioni quadrati!*" (lit., someone with square testicles, expressing highest respect).

** "I'm starving!"

How strong must the peer pressure among cosmologists be, if representatives of the Mediterranean way of life abandon their lunch in favor of a picnic consisting of fruits and vegetables!

But let's get back to the anomaly of the local Hubble law measured around our galaxy group. Once we had all taken a seat on the lawn, Roy Maartens opened the discussion, his lofty brow wrinkled. He spoke softly, but the others went silent, once he started. He asked a PhD student sitting next to him for his opinion of Wiltshire's lecture. The new blood realized that he was expected to display bared teeth at the dissident model, and retorted that he had found the main arguments unconvincing, in particular its theoretical background. Instantly the conversation veered into a highly mathematical discussion of differential geometry, such as the question how a fixed point in time could be represented in a four-dimensional space-time.

I was impatiently nibbling the raspberries, which had been passed around by Maartens, but like the others I didn't dare to interrupt the arguments about differential geometry floating over the picnic rug. All their busy debating was devoted to mathematically burying Wiltshire's idea. Like other alternative cosmologies, Wiltshire's attempt to challenge the standard model of cosmology may not have been exactly brilliant, nonetheless the anomaly of the local Hubble law he had pointed out contradicted the standard model. However, not a single word was said about this physical riddle during the entire lunch break, vanishing underneath the mathematical subtleties.

Close to our local group of galaxies, one of those vast empty spaces—dozens of millions of light years in diameter—begins. One great mystery is why these voids are so empty. We can find literally nothing there. Even James Peebles, a cosmologist not too far from the mainstream, confessed with a little irritation:

> I have been complaining about the absence of dwarfs in voids for at least 20 years…Does that mean I have been fooling myself? It's happened, but it is not what the evidence on void dwarfs continues to suggest to me.[15]

WHAT ISN'T ALLOWED TO BE, CAN'T BE—GALAXY CLUSTERS LARGER THAN PERMITTED

Cosmological structure formation was a big issue at the conference held in St. Petersburg in 2008. The first session was headed by Francesco Sylos Labini from Rome, with whom I subsequently spoke numerous times. Francesco's father was a high-ranking economic adviser in Italy's administration during the 1970s, until he resigned in protest against his government's involvement with the mafia.

Like his father, Francesco does not mince words. He analyzed the typical size of galaxy clusters with correlation functions, a technique of statistical physics in which he is an expert. In cosmology, we are assuming that the same mathematics holds true. Surprisingly, agglomerations of galaxies were even larger than previously assumed, some a billion light years across and more.[16]

We might even be tempted to think that the universe as a whole has an irregular shape, but shows the same typical structure at every scale, whether large or small, much like a fractal. As intriguing as this idea might be for some, for many it would be unbearable, since such large structures would contradict the standard model in a blatant fashion. According to established wisdom, the universe must become homogeneous on large scales, where the deviations from homogeneity are described by—don't laugh—an irregularity factor.

Francesco had done just a sober analysis, but such unwanted evidence doesn't bother a few leading research groups the slightest. The concordance model is their parade, richly decorated with many parameters, and they don't want it rained on. Toward the end of the conference, during a coffee break, I asked Francesco about his opinion on the new concepts of the cosmological standard model. He clasped his hands in the typical southern Italian manner and hit out: "If they don't even know how to count the galaxies properly, *che cazzo parlano della dark energy*?"*

STRONG RESULTS IN FAINT PHOTOGRAPHS

Cosmological structure formation shows a couple of contradictions, and some of these have been "explained" by a new parameter as modern cosmology continues to add adjustable numbers to its model. Astronomers with a certain overview, such as Mike Disney, have, however, pointed out that there are barely as many stand-alone observations.[17] Or, put a little more plainly, our belief that we have understood cosmology is merely a delusion.

Disney had, incidentally, predicted[18] low surface brightness (LSB) galaxies while analyzing the luminosity of galaxies in a groundbreaking paper in 1976. The Hubble Ultra Deep Field is a spectacular image of galaxies in the early universe, and it has turned LSBs once again into a hot topic. The early galaxies are, relative to their red shift, far too small, which poses a severe conflict with the Lambda-CDM model.[19] The numerous researchers

---

* Innocuous translation: "What the hell are they talking about dark energy?"

who deal with statistically bending the free parameters into shape usually do not take note of such irritating results.

> What is wanted is not the will-to-believe, but the wish to find out, which is its exact opposite.
>
> —*Bertrand Russell*

OUTSTANDING PROBLEMS, OUTLAW SCIENTISTS

It is highly annoying that such riddles are swept under the rug, solely because they do not fit into everybody's thought patterns. In 1970, Gerard de Vaucouleurs already weighed in, saying,[20] "I am increasingly concerned about the arbitrary rejection of evidence for the very reason that it contradicts our oversimplified models currently in fashion."

By now, the models of the universe have gotten more complicated and sophisticated, but new contradictions keep popping up anyway. It is hardly surprising that cosmology has many "outlaws." They are scientists who see the standard model as being fundamentally false, who then mostly present their own exotic, alternative cosmologies. There are, of course, a number of "outlaws," who are nothing more than blatherskites, but many are scientists with solid educations. Some of them even originate from within the astronomy establishment, including Fred Hoyle, Geoffrey Burbridge, Jayant Narlikar, Eric Lerner, or Halton Arp.

In 2004, the magazine *New Scientist* published an open letter signed by numerous researchers in which the Big Bang theory was called into question.[21] In my opinion, this was throwing out the baby with the bathwater, but it would nonetheless be foolish to exclude such voices or to gag them.

This raises a question. Why do some highly educated scientists draw such drastically different conclusions from the same observations? At least, one should check their claimed contradictions to the standard model. And we should not be surprised when they are occasionally mistaken, or pursue their own model all too wide-eyed. After all, it is difficult not to sink while fighting an armada of researchers of the standard model. There is a type of the cool overachiever you can spot after listening to the first minutes of his talk: the problem serves just as a stage for the solution, adorned with colorful animations, where the consistency with the Lambda-CDM model is as safe as the Bank of England, and the cocksure attitude stalls only when, after a critical question, the thesis adviser has to come to the rescue.

Alternative cosmologies sometimes sound insane, but we should take it seriously when they reveal the weak points of the accepted notions. The Big Bang, as such, is barely a quantifiable theory. Instead, we have merely had a guess of a possibly appropriate picture of the universe's early days, which in principle gets along with a whole number of gravitational theories. Hence, I wouldn't discuss the Big Bang all that much, but rather call into question whether the detailed story we are being told is true. It is entirely based on the standard Lambda-CDM model, which rests upon general relativity. However beautiful Einstein's theory may be, it is utterly naïve to extrapolate a theory to the entire universe when it has been tested only in our solar system.

A NOW LONELY HERETIC

It is mostly unorthodox thinkers isolated from the scientific community who have achieved major breakthroughs—at least, that is what the Italian science historian Federico di Trocchio says. But how do we distinguish the visionaries from the crackpots? Di Trocchio sees Halton Arp as a modern example of a "heretic," an excellent cosmologist who was awarded the Helen B. Warner prize in 1960 as a young and promising astronomer.

However, in the 1970s, he was at odds with the cosmological mainstream and had his telescope time cut. After that, he moved to the Max Planck Institute near Munich. I was intrigued, and in 2007 I asked him if we could meet, and he kindly agreed. Arp is a lean man with bright keen eyes. Only a subtle trembling of his hands betrayed his 80 years, and his well-dressed appearance stood out next to scientists' usual lumberjack shirts.

Although I was fairly unfamiliar with his theses, he patiently explained quasar phenomena to me and advised me on several topics one should reflect upon. He spoke calmly, without the evangelical fervor of eccentrics wanting to persuade you, and sometimes seemed weary. Other researchers agree that the nature of quasars, due to the various contradictions, is not as clear as it is made out to be (As it was recently pointed out in an excellent article by the Spanish astronomer Martín López Corredoira.[22]). I do think that the theory Arp favors must be erroneous, but I was nonetheless impressed by his knowledge and his intelligent reasoning.

Later, I read his controversial book *Seeing Red*. Even if some claims in it might not withstand the observational test, Arp raises questions that are usually ignored by the conventional fashion. Such a perspective should be a lesson to every astronomer. There is almost no unique interpretation of data, and we should take care, especially when one opinion has been generally accepted.

Regrettably, Arp's depth of experience isn't appreciated too much at the Institute. We had our conversation in a little office instead of in the cafeteria, because the younger scientists, he said, are a little uncomfortable being seen with an outsider. It seems that even the simple exposure to alternative theses is regarded as a risk of infection in some places. Arp didn't want to create difficulties for the young scientists, he said considerately, without any sign of bitterness. A fine old gentleman.

MODERN COSMOLOGY—EPICYCLES AS FACTUAL NECESSITY

At a lecture about gravitational lenses some years ago, I recognized my old college friend Matthias Bartelmann, by now one of the leading cosmologists in Germany. We knew each other from the first year in physics, and I had been especially impressed that he had already published a textbook of astronomy back then as a 19-year-old. We agreed to have dinner at an Italian restaurant, but since we hadn't seen each other for almost 20 years, I began the physics conversation carefully.

"You consider yourself a mainstream cosmologist, right?" I asked and continued after his nod. "Do these observations really fit together so well?"

He said he sees quite a good agreement with the standard model, though admitting that it rests upon many assumptions that build upon each other. As he indulged himself in his walnut pesto, I addressed another controversial issue.

"General relativity is admittedly very neat. But does that precisely mean that it is valid? In medieval times, people had found the planets' orbits neat and thus artificially upheld the geocentric world view. Couldn't it be that we are following the same complicated road like that of epicycles?"

To my surprise, he didn't object. "But what should we do otherwise?" he responded. "Until some genius comes along and understands everything, we have to continue working with these models."

While I was looking forward to a relaxed evening, Matthias had to rush to a meeting on "The Dark Universe" by the Excellence Cluster, a cutting-edge research initiative. Such large-scale projects are favored by politics as it brings together the elites. In practice, however, the scientific diversity of opinions is homogenized. Supposedly, such events are more harmful than useful for intelligent people.

Before Matthias left, and after a glass of wine, I dared to probe even further. "Matthias," I said, "tell me frankly, what do you think about stuff like string's theory inflation on a brane? After all, there isn't the slightest chance of ever testing it. Why does nobody complain about this bullshit?"

"I don't know what to do with it either," he replied and leaned forward with a little chuckle, "but I know many very bright people who devote their time to this matter."

Yes, of course. After all, we are only humans! Writing this, I realize that I am struggling between sympathy for Matthias and the obvious fact that he, too, has slid into the cosmological groupthink. And probably, it would be hard for me to have a friendly dinner with those very bright people while telling them what I think about their inflation on a brane.

WHERE DO THE NEW NUMBERS LEAD AFTER ALL?

Astonishingly, many astrophysicists admit that the current period in understanding the science of the universe reminds us of epicycles. They realize that recent complications, such as a temporal variation of dark energy or a new interaction between dark matter and dark energy,[23] just make the model messier.

Nevertheless, too many scientists constrain themselves to pure data collection and allow their observations to be interpreted in terms of the standard model. True, Galileo was an observer, too, and it may be that Copernicus would have sunk into historical oblivion without him. But is it absolutely necessary to stick to a single model—especially if it has become as messy as the geocentric world view?

Today, we have to question more than ever whether the law of gravitation is valid outside of the solar system. It is naïve to make that extrapolation, the observations have been full of contradictions, and the numerous obscure assumptions indicate a crisis.

Many years ago, Alan Sandage made that point when stating that cosmology is the search for two numbers. He meant Hubble's constant and matter density. By now we have reached seven or eight, and we are collecting more and more accurate data. But is this leading to true progress in our understanding of nature? With all the new concepts in cosmology, the following words attributed to Richard Feynman keep coming to my mind:

> Correct theories of physics are perfect things, and a replacement theory must be a new perfect thing, not an imperfection added onto an old perfect thing. This is the essence of "revolution," the replacement of the old with a new, not the adding of more crap onto the old.

This bluntly summarizes what the science philosopher Thomas Kuhn had theorized as a "paradigm shift," which necessarily happens after an increasingly complicated model is accompanied by a pile-up of anomalies. We are in the midst of such a crisis, and a scientific revolution is yet to come.

Chapter 8

PRECISION IN THE TEA LEAVES

MESSAGE FROM THE COSMIC MICROWAVE BACKGROUND: HOW MUCH IS JUST NOISE?

It gives me the creeps if I imagine that the universe was once just hot plasma, composed of electrons, protons, and a few helium nuclei. Right after the Big Bang, the particles in the plasma were still moving too fast in all the heat and excitement to feel the electrical attraction between them. But the universe's expansion inevitably resulted in everything cooling. Once the electrical attraction began to dominate, the particles started partnering up into neutral hydrogen atoms (one proton, one electron), and helium atoms (two protons, two neutrons, and two electrons).

This drastically changed the situation for electromagnetic waves, and thus light. In the plasma, light was deflected by every single particle. But once the charged particles paired up, light "decoupled" from the neutral matter and could pass through it without any bother. All at once, the bright universe became invisible, as if someone had snuffed out a candle, and the temperature sank to about 3,000 Kelvin.*

Our eyes would still perceive this as a mild reddish glow, except that the universe's expansion since then has stretched light waves by a factor of 1100—what's called redshift z—into a range we now know as "microwaves."

---

* Temperature is just another name for the kinetic energy of atoms. At 0 Kelvin, corresponding to –273°C, all particles are practically at rest.

That's the origin of the name "cosmic microwave background" (CMB for short), which today shows a radiation temperature now cooled by the same factor of 1100 from a temperature of 3000 Kelvin to about 2.7 Kelvin. This scenario is believed to be the explanation for the radiation so successfully recorded by the COBE and WMAP spacecraft. And now we admire the data from the even more precise and more sensitive Planck mission. What do we learn from this?

AND THE UNIVERSE ISN'T MOVING AFTER ALL

It's easy to recognize the motion of Earth around the Sun in the CMB data. The Doppler shift apparently causes seasonal red and blue shifts of the respective sky regions. Furthermore, by analyzing the data, we can "see" the solar system orbiting the Milky Way's center, as well as a proper motion of the Milky Way toward a huge mass concentration of a galaxy cluster.

Interestingly, the cosmic microwave background additionally reveals that neither the Hubble expansion of the universe nor the Big Bang itself should be mistaken for an explosion, for we detect the radiation with astonishing intensity. If we were to look upon receding gas clouds of the early universe, the Doppler shift would cause the very same decrease in intensity by a factor of 1100. Therefore the expansion of the universe does not mean its objects are flying apart, but rather that they are getting farther apart as a result of space itself expanding—a concept of general relativity that raises many questions.[1]

The CMB data reveal another characteristic of the cosmos that is quite seldom mentioned, maybe because theorists don't have much desire for it. One of Galileo's big insights was that the laws of Nature do not depend on whether we are in uniform motion or at rest. As you may remember, Einstein discovered his theory of relativity precisely because no experiment, not even measuring the speed of light, could determine uniform motion.

As of recently, we can observe such uniform motion: the signals of the cosmic microwave background clearly show that we are not at rest but rather moving at a speed of 230 miles per second toward the Crater constellation, which is actually quite slow compared to the speed of light. Whether this will one day be important for theoretical physics is an open question, but it is a fact that the cosmic microwave background defines an absolute frame of reference. For the first time, we know what "rest" means. However, the theories of Galileo all the way up to Einstein have told us that it doesn't matter.

DESPERATELY SEEKING FLUCTUATIONS—AND
AN OBSCURE FINDING

The first all-sky measurement of the cosmic background radiation from the COBE satellite in 1989 brought a couple of surprises. The radiation was unbelievably homogeneous, a fact that caused some headaches for the researchers who were analyzing the data. It was simply too perfect! The seemingly uniform temperature everywhere in the early cosmos was somehow beautiful, but also puzzling. Slight irregularities, referred to as fluctuations or anisotropies, would have been necessary to explain the subsequent development of galaxies.

After almost two years of intense efforts scrutinizing the raw data, the COBE team found the looked-for temperature fluctuations. But they were much smaller than expected, in the range of a few millionths of 1 Kelvin, utterly insufficient to account for galaxy formation.

Help came—as it so often does—from dark matter. The pragmatic idea is as follows: if we can't see the required large fluctuations in normal matter, then these have to have existed in dark matter. Based on this idea, dark matter rushed ahead of normal matter, massed itself together, and formed the invisible basins of gravitational attraction where the lethargic normal matter eventually accreted.

Many cosmologists see this as "independent evidence" for the existence of dark matter. We could all be happier about this if this wasn't once again, to put it bluntly, a fudge factor. Modern physicists seem to have few qualms about resorting to such devices.

Now savor the following reasoning. How strong do these fluctuations of dark matter have to have been to "seed" galaxy formation? The answer is obvious—precisely as strong as the ones that produce the distribution we now see. We have, in this case, very accurately "measured' the fluctuations. It is nice to have dark matter come to the rescue again, but it is becoming rather like we are talking about a criminal on the loose upon whom we can pin as many offenses as we wish.* As long as we don't have any idea of what these invisible 'dark' particles are or how they behave beyond being a source of gravity, they are free to be the cause of everything and anything in cosmology.

> Once the problem is remediated by an excuse, there is no need any more to reflect upon it.
>
> —*Erwin Schrödinger, Nobel laureate 1933*

---

* We did not deepen other dark matter occurrences as gravitational lensing and galaxy clusters.

DARK TRACES IN THE CMB

The latest CMB research now focuses on the patterns that form on the images of the sky, which are the typical size of the spots illustrating temperature differences (fig. 9, left). It turns out that the fluctuations with an approximate angular size of one degree are the most pronounced ones in the celestial sphere. Let's think back to the time when these signals emerged. Back then, the visible universe was a lot younger and a lot smaller because light hadn't had much time to spread. One angular degree correlates to about the size of the horizon of the visible universe at that time.

The primordial plasma soup of hot matter and radiation was a medium in which sound waves (which are also pressure, and thereby temperature waves) could oscillate. In this manner, the waves were able to leave their traces in the microwave background, indicating the size of the universe when the plasma era ended.

Researchers are looking for the fluctuation patterns based on wave harmonics, which produce identifiable wave patterns that leave still smaller spots branded in the sky, again representing temperature differences. The second wave harmonics, producing a different wave pattern, are supposed to mark the presence of dark matter—assuming the current model to be correct. The measurements, however, do not always agree with each other.[2] Moreover, the "dark matter" wave amplitude depends on almost every cosmological parameter—and there are many.

Inferences of this kind, in combination with the hypothesized effects of dark energy, are the highlights in lectures on "precision cosmology." Scientists happily display charts with colorful ellipses, each one representing a data set like the CMB. If the ellipses' positions, tuned by a couple of parameters, more or less overlap, it's trumpeted as a triumph of the standard model.

Unfortunately, many scientists forget how many assumptions this model rests upon. Though the measurements are not any more contradictory than they were before 1998, the wonderful self-consistency was dearly paid for through a new parameter called dark energy. And none of these neat diagrams explains the many inconsistencies of dark matter. A friend of mine working in a group that deals with dark matter tests in the solar system wrote to me:

> If you assume that dark matter has this and that fraction, that there is a cosmological constant, that fluctuations are adiabatic,* that neutrinos have a certain mass and so on, at the end you will surely "find" that the deviations from Newton's law are almost zero... but if you want to scrutinize something properly, you're simply not supposed to start out of such assumptions.

---

* Without heat exchange, like deflating a bike tube.

A COSMOLOGICAL CONCERT WITH THE WRONG SHEET MUSIC?

Cosmologists assume that the cosmic microwave background originated from a period 380,000 years after the Big Bang. But what followed? The typical size of the fluctuations of density back then should have manifested itself in the later universe as the typical size of galactic structures. The obvious place to look for evidence of this was to search through the data from the Sloan Digital Sky Survey on 700,000 galaxies. And there it was. In 2005, Daniel Eisenstein and his team at the Harvard-Smithsonian Center for Astrophysics looked for and found a small deviation indicating such primordial density fluctuations, technically called "baryonic acoustic oscillations."

However, the waves radiating in spherical ripples from the plasma period out through the cacophony of the universe are not easy to track. As Eisenstein put it, "The Universe is composed of many over-dense regions whose spherical waves have overlapped. If one region is like throwing a pebble in a pond and seeing the expanding ripple, then the Universe is like throwing a handful of gravel in a pool."[3]

Once the group had found the signal in the desired magnitude corresponding to the CMB data, all the diagrams summarizing the cosmological measurements were soon decorated with an ellipse in a new color, well suited to the others.*

In December 2008, the last data gaps of the Sloan Sky Survey catalog were filled in so that researchers could call upon the data for all known galaxy positions of one-quarter of the sky. Around that time, I spoke to Francesco Sylos Labini from Rome. He was upset. He had just completed filling his statistical programs with the newest data, and where the lauded acoustic peak had been found in 2005, he had found…nothing.[4] I can't imagine anyone doubting Sylos Labini's competence and carefulness. Does this mean that the colorful diagrams that are going around the cosmology conferences throughout the world are wrong?[5] I do not know.

Incidentally, this is a situation science sees itself confronted with quite often. Perfectly respectable scientific teams come up with differing results, sometimes plainly contradicting each other. This does not mean that scientific adversaries are using wrong methods, or that there is any intention to deceive. Scientists are human beings with their own biases, usually in favor of their preferred theories. The problem is not so much that there are conflicting views as it is about the way conflicts are normally handled. The more accredited researchers will typically have their results published in the more prestigious

---

* Interestingly, another group member who had later repeated the analysis, found a lower statistical significance for the sensational effect, quite unusual. See subsequent endnotes.

journals, and hence their preferred interpretation will most often prevail. In cases such as the galaxy distribution, however, it would be much more helpful for results to be made transparent by putting the raw data and the computer code online, and open it up to public evaluation. Until then, I suspect, everyone will continue to go with the baryonic acoustic oscillations.

THE SOLAR SYSTEM: A STING IN THE TALE OF THE COSMOS

Let us return to the microwave background. Besides the tiny spots, there are also large-scale spatial temperature fluctuations, called quadrupole and octupole moments, illustrated in fig. 9 on the right. (Dipoles have two poles; octupoles have eight.) It is as if the sky was a gigantic drum with some overtones more audible than others. The unusually pronounced octupole divides the celestial sphere into eight pieces through nodal lines. Where can such a pattern come from?

An obvious guess would be that the Milky Way had a hand in it, since countless stars and gas clouds are lined up at this belt on the night sky, and all of this foreground noise has to be filtered out of the data. Thus, it wouldn't be a big surprise if, say, the nodal line of the octupole coincided with the belt, a possibly inevitable error due to incomplete filtering. But where is this nodal line actually located? It is precisely in the plane in which planets of our solar system orbit, the ecliptic. But the ecliptic is completely different from the Milky Way's plane.

This is a very odd result, known as "octupole anomaly." Should our humble solar system be branded into the universe's face? If we don't resort to the esoteric, the only reasonable explanation is some subtle error in the analysis related to the framework of the solar system. Irritatingly, the anomaly seems to persist in the new Planck data, and a related anomaly, called hemispheric asymmetry, showed up.[6]

In general, evaluating data from the mapping of the background radiation, such as that produced by WMAP, is a highly complex endeavor, and more transparency would be desirable. For example, initially the team refused to publish how the foreground signal of the Milky Way was subtracted from the data because of "complex noise properties."[7] But other aspects of the method have raised questions from unusual quarters.

Pierre-Marie Robitaille, a professor of radiology at Ohio State University, is a pioneer and expert in magnetic resonance imaging. He has applied his talents to astrophysics, turning a critical eye on the CMB data collection. He was particularly critical of what he considered the rather crude methods used to subtract the foreground noise from Earth, the solar system, and the Milky Way, which is about 1,000 times stronger than the signal

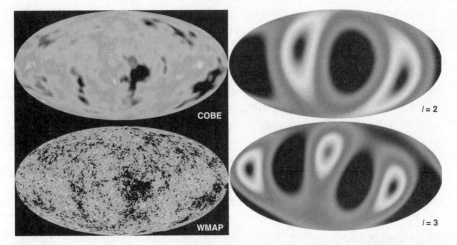

Fig. 9. *Left*: Temperature fluctuations in the cosmic microwave background, recorded by COBE and WMAP. *Right*: Quadrupole (l=2) and octupole (l=3) anomalies of the temperature distribution, extracted from the WMAP data. Notably, their orientations coincide and at the same time, the plane of the solar system appears as one of the nodal lines. Planck's data are still more precise.

researchers were looking for. According to Robitaille, lab experience in medicine demonstrates that it is impossible to extract a signal that is so many times smaller than the background without the risk of producing artificial effects.[8] He says that WMAP and COBE scientists were simply trying too hard to squeeze good results out of the data.

It is certainly too much here to recall Mark Twain's dictum: "Get your facts first, and then you can distort them as much as you please." But again, publicly accessible and well-documented computer code would create more confidence in the cosmic microwave maps released by WMAP and Planck. Slowly, things seem to change. The Chinese researchers Hao Liu and Ti-Pei Li published the code of their reanalysis of the microwave background. They have found serious contradictions to the standard model.[9]

LIVING IN THE SPACE BETWEEN MATH AND PHYSICS

Mathematical physicists are generally pleasant people. I once attended a little conference about continuum mechanics in Udine, a lovely town in northern Italy, where I had the opportunity to get to know a few of them. There was the Irish professor, wearing green socks and sandals, who was juggling with ellipsoid transformations, as if these easily fit into his big head. Young couples on dates in the *piazza* gazed in amazement at two researchers from Milan, as they (around midnight) vociferously argued about quaternion groups, a four-dimensional generalization of complex numbers. One mathematician from Verona, excited about category theory, forgot to eat during dinner, while his listeners dined attentively, if uncomprehendingly.

As strange as some mathematical physicists may seem, they are, generally speaking, in a comfortable position. They have come a long way since the early 1900s, when mathematicians David Hilbert, Felix Klein, and Max Born turned physicists loose on the canons of mathematics that are too often unrestrained by physical reality. Among mathematicians, the work of mathematical physicists is regarded as useful and application oriented. Among physicists, they are given *carte blanche* to do things out of touch with reality. The mathematical physicist usually gives great thought to matters that most people take for granted, but is never blamed if his ideas don't work.

When such mathematical physicists are let loose on cosmology, they start to speculate about the topology (shape) of the universe as a whole. Maybe the universe has the form of a bicycle tube, or is shaped like a multiply cut-and-glued hyperpretzel. Why not? In any case, we can all do beautiful calculations. Dozens of papers from mathematical physics institutes have been dedicated to the idea that such nontrivial topologies could cause an octupole anomaly in the cosmic microwave background. Math was important for the cosmos! The explanation of the real riddle—why this odd orientation is related to the solar system—has been left for the rank-and-file astronomers.

THE IRON CURTAIN: LIGHT AND MATTER
SEPARATED FOREVER

The European Space Agency's (ESA's) Planck spacecraft is a step up from the American COBE and WMAP in detailing the cosmic background microwave radiation. Enthusiasts of the cosmic radiation and skeptics, for whom the WMAP data analysis left too many open questions, agree on the mission's scientific value. Future detectors, however, will barely improve the Planck results, because all that annoying noise originating from the dust belt of the Milky Way cannot be polished away any more. A speaker at the conference "Open Questions in Cosmology: The First Billion Years" in Germany in 2005 put it in a nutshell: "The limit of background measurement is foreground."

Although this statement is not exactly sensational, it seems to sink in very slowly. Sure, the data as such might be of exceptional accuracy, so that professional enthusiasts like Joel Primack have coined the phrase "precision cosmology" for it. Others call the parameters of the standard cosmological model measured through the background radiation the "gospel according to WMAP."

We cannot lose sight of the fact that the light of the background radiation was exposed to many highwaymen on its 14-billion-year journey to our sophisticated detectors, and this compromises the accuracy of the data. There are so many ways that the wavelength of light might be altered by gravitational fields of galaxy clusters, collisions with free electrons, and scattering processes we don't understand exactly. And that doesn't even include the kind of curve ball dark matter might throw at us at any time. Even if we succeed in eliminating all foreground noise, the data still depends on the cosmological model. So, despite the neat precision, there are just too many factors we can't get out of the meshwork.

The cosmic microwave background allows us to look as far back in time as we can. Consider the enormous redshift of $z = 1,100$ compared to that of distant galaxies or quasars, where $z = 10$. But the universe's opacity for the first 380,000 years is an insurmountable barrier future cosmologists will have to live with. WMAP is sometimes said to have produced baby pictures of the universe, but wouldn't it be nice if we could get something equivalent to a first-trimester ultrasound picture? We can't, so the view prior to the end of the plasma era is out of reach. As a consequence, in understandable curiosity, some want to extract more out of the data that we do have. A widespread bad habit in science.

LANTERNFISH SPECIES CLEARLY IDENTIFIED

If an electron hits its rare, exotic, mirror-image positron, both end their lives in a dramatic gamma-ray flash of pair annihilation. For that reason, particles such as the positron are called "antimatter." Conversely, particle couples made of matter and antimatter can be generated out of a single photon just as well. Even the much heavier proton-antiproton pairs can be created if, of course, the photon's energy is high enough to meet Einstein's formula $E = mc^2$. Antimatter puzzles physicists a lot. It hardly ever appears in everyday life because it would promptly be annihilated by its counterpart in normal matter, but it is always present when new particles are created.

The heavier the particles are the more effort is needed in big colliders to produce them, so it is sometimes speculated that these particles could have been formed cost-free shortly after the Big Bang, in the so-called primordial phase of tremendous heat. It's a wonderful idea, but unfortunately it is also untestable. How could the information about high-energy particle creation and annihilation survive for 380,000 years in a sizzling hot soup? Notwithstanding this, I have heard conference presentations in which people in all seriousness talk about their hope to detect such "signatures" of pair annihilation in the cosmic microwave background. This is already absurd because of the complexity of what happens to radiation after it is released at the end of the plasma era, but especially because we know literally nothing about what happened prior to that. How many physical processes could have superimposed themselves on the radiation and rendered any "signature" void by this point!

You can compare such a data analysis to that of satellite images of the ocean surface. We can ascertain the sea level to the centimeter, and infrared images can tell us the exact temperature. Thus, one can clearly identify the gulf stream of the North Atlantic, maybe even deduce the salinity and the algae percentage from a spectral analysis. And if you're lucky, you read the wind speed from the ripples on the water. But finding primordial particles on the WMAP chart would be as if, by analyzing the movement of the ocean surface, you would identify deep sea fish and classify them zoologically. This doesn't mean, of course, that there aren't several research groups devoted to doing exactly this.

> Who does not lose one's mind while reasoning about certain things, has nothing to lose.
>
> —*Gotthold Ephraim Lessing, German writer*

PRECISION CHEMISTRY RIGHT AFTER THE BIG BANG?

The cosmic microwave background is the earliest information we can get about the universe at different wavelengths and with a high angular resolution. Is there actually a method to observe even earlier periods? We have to go back a little in history.

About 200 years ago, the Bavarian physicist Joseph Fraunhofer discovered fine dark lines in images of the solar spectrum, little gaps in the continuous color spectrum of light emitted by the sun. This happens when, for instance, helium atoms are located on the sun's surface and absorb light of "their" wavelength as the rest of the wavelengths continue on their path to Earth. The less of the helium wavelength that comes through, the higher the concentration of helium is on the sun's surface.

An analogous method, with modern telescopes, can be used to measure the helium and hydrogen concentration in remote gas clouds in the early universe, though we have to assume here that their chemical composition has gone unchanged since then. In this case, it is not the sun but quasars, which already existed at the time, serving as the light source from behind the gas clouds. It turns out that even back then—before any star had formed—about 25 percent of the universe's mass consisted of helium nuclei. This is remarkable, because the activity of stars (all of the sun's power comes from fusion of hydrogen to helium) could not form such an amount.

We can also detect minor quantities of deuterium (hydrogen with an extra neutron), but not to a good accuracy.* To summarize, the light elements in the early universe are interesting, but their measurement is of limited precision. Nevertheless, cosmologists hope to get with it a picture of what happened before the formation of the cosmic microwave background—that plasma curtain hindering any direct observation. What else do they need to look behind?

It is theoretically assumed that the same number of protons and neutrons existed at the very beginning. Now a cruel selection begins. The neutrons decay into protons, electrons, and antineutrinos on average every 15 minutes, unless the neutrons manage to partner up with a proton and form a (stable) deuterium nucleus. The clock is ticking and the universe is expanding and it is becoming more unlikely by the second for a neutron to catch a proton. Deuterium can then fuse into helium, but it doesn't have much time either. And there is a little lithium (the third element), though the amount doesn't fit the model.[10]

As a result of this dance, only about one-seventh of the neutrons survived, and most of them did so as helium constituents. If one further assumes that the half-life of neutrons and various other properties of the early universe's matter have also remained unchanged until today—somewhat daring as it is—then one can even create a model of the density back then and get some idea of the expansion velocity of the universe.

Steven Weinberg described this story of primordial nucleosynthesis in his book *The First Three Minutes*, popularizing an interesting topic. But in doing so, he rather screwed up a little the morals of how to do reliable science. The story of the first few minutes rests upon too many theoretical guesses. As the physicist Roger Penrose has pointed out, such models naively presuppose that everything began in thermodynamic equilibrium, which is probably completely false.

There is rarely a more hypothetical mix of numerous crude assumptions, indirect conclusions, and huge measurement errors than is used to describe the primeval soup. Primordial nucleosynthesis provides a qualitative picture and nothing more. It still leaves aside the question of whether the laws of Nature were really identical to what they are today and whether our perception of time is correct. We don't know.

---

* Since no quasar has both helium and deuterium lines, direct comparison is impossible.

The wise man doubteth often, and changeth his mind; the fool is obstinate, and doubteth not; he knoweth all things but his own ignorance.

—*Akhenaten, Egyptian pharaoh*

SUPERNOVA 1987A—THE PARTY EVERYONE WANTED TO JOIN

The idea of primordial nucleosynthesis does sound quite plausible. However, the quantifiable results are, among other things, subject to the number and mass of the antineutrinos originating from nuclear reactions. The importance of neutrinos for astrophysics suddenly became clear at an event on February 24, 1987. "Shelton 19871 70224 333// 05343 16916" was the message to the International Astronomical Union, sent in a telegraph by the astronomer Ian Shelton from the Las Campanas Observatory in Chile. The cryptic sequence in the message contained the coordinates of a sensation—the brightest supernova explosion since the one observed by Kepler in 1604!

Although 1987 was just before the Internet era, the news spread like wildfire among astronomers, and everybody quickly focused their telescopes on the Large Magellanic Cloud, an accompanying galaxy of our Milky Way, only 160,000 light years away. Physicists at the big neutrino observatories around the world figured that the starburst must have caused nuclear reactions and released a wave of neutrinos that would have already passed through Earth but might have left a trace. Immediately, a fevered search for the neutrinos from the supernova began. Researchers combed through their data and found the looked-for signal.

Scientists at the Italian-French neutrino observatory in Mount Blanc detected a neutrino shower on February 23, 1987, at 2:52 CET, while a US lab detected an extraordinary signal at 7:35! A Russian observatory reported extra neutrinos at 7:36, although Russia isn't exactly a light minute distant from the United States. The most interesting time measurement came from the Kamiokande observatory in Japan. Unfortunately, their clock lacked synchronization, but people remembered it displaying the time as 7:35.[11]

WISHFUL THINKING BEFORE THE COSMIC MICROWAVE BACKGROUND

Neutrinos are notoriously difficult to detect because they tend to pass through everything in their path without interacting with any of it. But once in a while, one of the huge detectors set deep in worked-out or abandoned mines manages to snag a few. The team at the Kamiokande laboratory, headed up by Masatoshi Koshiba, counted 11 hits out of the 10 quadrillions of neutrinos that had passed the detector. Those glorious 11 were enough for Koshiba to win the Nobel Prize in 2002. Detecting neutrinos from another galaxy is something exciting, but it would have been nice if the result was independently confirmed in same way that WMAP corroborated data from COBE.

The Americans, Russians, and Japanese researchers agreed that their times for the neutrino detection were more or less the same, which caused considerable embarrassment for the Mont Blanc collaboration. This conflict over the data triggered a number of hostile debates at conferences, and allegedly, some of the authors of the Mont Blanc article wanted to have their names crossed off the publication. Mistakes can occur, but we have to ask ourselves, what would have happened if, back then, only the Mont Blanc observatory was operating? Would the Nobel Prize have then been awarded for something that is now considered a statistical flaw?

Although a generally accepted picture exists of the way neutrinos should behave, quantitative measurements in astrophysics still fight with difficulties (and with each other). It is easy to see the appeal of the promise that the plasma "curtain" of the microwave background will someday be transparent for neutrino "telescopes" and gratify us with data about the early universe. This is boundlessly optimistic, since we have so much trouble identifying neutrinos from right next door.

Similarly, the hope of seeing gravitational waves from the primordial era flourishes. This may be possible in principle, but the detectors have been quite unresponsive to such hopes. Success remains wishful thinking, as long as we cannot even measure the gravitational waves of our nearest neighbors in the galaxy.

MEASURING FOR THE SAKE OF MEASURING

Scientists and science funding agencies have invested a great deal of time, talent, and money into high-tech observational equipment, here on Earth and out in space. But as fascinating as modern observation techniques may be, the hunt for accuracy becomes a sport rather than a science. It is the understanding of the data we are missing.

It may be exciting, for instance, to be working on "the faintest objects ever observed," but chasing weak signals does not necessarily mark the frontline of research. The astronomer Mike Disney noted, with some acidity,[12]

A major fraction of the prime time on all large telescopes has been devoted to the study of objects right at the horizon, with, or so it seems to me, very little result. To be rude about it, statistical studies of faint objects can keep a career going for ages without the need for a single original thought.

A bigger problem with studying early universe faint objects is that one needs to subtract a lot of unwanted foreground signal—stars, molecular clouds, etc.—and there are many sources of error. Not everyone is familiar with the pitfalls that lie in image processing. For instance, while planning a map of the 21 cm hydrogen line in the Milky Way, scientists seemed confident that radio noise 10,000 times stronger than the signal could be properly filtered.[13] This doesn't mean that they are unqualified or even disingenuous. But overestimation of accuracy is a general problem in physics. Admitting the possibility of systematic errors (from unknown effects) in the data easily damages one's reputation. And announcements of discoveries, even if too optimistic, are always more welcome than scientifically honest statements that one failed to properly eliminate the noise.

The further we go back in time, the greater is our desire to look at the exciting era right after the Big Bang. However, the prospect of getting reliable knowledge about neutrinos, gravitational waves, and the chemical composition is still poor. Even the celebrated cosmic microwave background gives little more than a rudimentary glimpse of the early universe.

> If the scientists didn't try, they were bound to fail. But the harder they tried, the more sure they made themselves the playthings of history.
>
> —Harry Collins, science historian[14]

Part III

DEAD END

Chapter 9

MUDDY WATER

THE COSMOLOGY OF DARK PIXELS IN THE FIRST DARK AGE: GIVING WORK TO SUPERCOMPUTERS

On the computer screen, a colorful flash bursts out of the darkness, and a sea of bright dots covers the screen. The dots immediately start organizing themselves, seemingly at random. A clock is running at the edge of the screen. Millions of years are going by in seconds as our universe's first epoch is packaged in a cosmic time lapse. As the clock is rapidly counting out billions of years, fascinating structures are evolving into filaments of galaxies, still seemingly random. But there is nothing random about this. It has all been precisely calculated by the Millennium Simulation,[1] one of the most ambitious projects yet for numerical astrophysics.

The simulation, run by an international collaboration, modeled the gravitational evolution of a section of the universe 2.2 billion light years across. Despite having the enormous computing power with a capacity of 1 trillion operations per second, the machines were kept busy for a whole month with terabytes of data. The resulting high-speed evolutionary picture is meant for the movie screen, and not surprisingly, *Nature* dedicated a cover to this simulation of the universe in 2005. The silicon technology is impressive and the programming work is worthy of pride. Anyone who has ever worked with the pitfalls of simulations and written source codes knows the satisfaction of the computer finally doing what one want it to do, and then delivering beautiful results. What do we learn from this after all?

PUBLICIZING WITH PICTURES OR SCIENCE WITH NUMBERS?

The Millennium Simulation showed that quasars could have emerged in the early universe. Fine. Physics, however, is a quantitative science. What if we want to know concrete numbers? We observe about 30 satellite galaxies of the Milky Way, while the simulation produces approximately 1,000 of these little galactic companions. New results show an even larger number. According to the simulations, the night sky would be lit up by all these satellite galaxies.

Evaluating numerical results means talking about numbers, and not just making statements about how nicely the pictures match the early universe. But then the results often appear a little less convincing. A lot more disconcerting is the fact that computer models generously help themselves to a series of free parameters. One parameter is the so-called "bias" in galaxy formation, which describes how much more structure the early universe contained than we can understand. The more fudge factors you invoke in your simulations, the nicer the outcome fits reality. But this is not a prediction.

At the very least, this epistemological bad habit of incorporating free parameters (which is evocative of epicycles) should be documented. But most simulations don't reveal to what degree they are stuffed with these squishy things. Robert Sanders writes in his book *The Dark Matter Problem*:

> [This]...has led to the industry of "semi-analytic galaxy-formation" modeling. Here, poorly understood aspects of the dissipational component of galaxies—effects such as gas cooling, star formation and supernovae hydrodynamics—are modeled by simple analytic relations characterized by a number of adjustable parameters: supernova feedback, timescale for star formation...I count as many as eight such free parameters [...] When the observations are matched, the model is claimed to be successful.[2]

95 PERCENT OF THE PROBLEMS SOLVED THROUGH 95 PERCENT DARK SUBSTANCE

Despite the brute force computer power of the Millennium Simulation, researchers had to settle for a somewhat lumpy grid to model the universe, such that one pixel represents a billion sunlike stars. Normal matter evidently starts forming structures at one solar mass, so this produces a rather clumsy model. Dark matter on the other hand, with its opaque properties, made the simulation work very nicely. The usual normal (baryonic) matter was folded into the model at a second step, as we learn from the project website. Was it just a nuisance before?

The well-known astrophysicist Jerry Ostriker reported at a meeting in 2007 that one should do the simulations completely without the baryonic matter—in other words, just leave normal matter out of the model of the universe. Think about this for a moment. Thanks to the computationally well-behaved dark substances, soon the last detail problem will have been eliminated: reality. But how can we model our universe if we base it on dark matter, about which we know nothing but its gravitational effects? What if dark matter turns out to be a delusion because, say, the law of gravitation is false? That would render the simulations of the universe obsolete. It would have been nothing more than dust thrown into our eyes.

THE COMPUTER AND ITS DOTTY INDEPENDENT LIFE

Attempts to simulate dark matter are almost as old as the computer itself. Many hypotheses for dark matter have come and gone since then. The initially favored neutrinos went out of fashion because they have not yet shown themselves to have a measurable mass. But computing capacity has gotten larger and larger, and there is an enormous temptation to make your research fit with what a computer can do rather than make the computer programming fit your research. The need to make use of these computing beasts can too easily drown the creative thought process.

Can we really understand the universe by means of the numerically cheap dark matter and a lot of computing power? Frank Wilczek, who was awarded the Nobel Prize in 2004 for his work in theoretical nuclear physics, says,

> Fortunately, our near-total ignorance concerning the nature of most of the mass of the Universe does not bar us from modeling the evolution of its density. That's because the dominant interaction on large scales is gravity, and gravity does not care about details.[3]

In contrast, Mike Disney from Cardiff University, who is possibly slightly more familiar with astronomy than Wilczek, looks at the "details" differently:

> They start off with a whole lot of cold dark matter "dots," the dots apparently form filaments under the force of gravity...and we are supposed to admire the result. What result? That to me is the question. Presumably we are supposed to compare the dots with real structures and infer some properties of the physical Universe. In my opinion it is nothing more than a seductive but futile computer game.[4]

Further, Disney describes this "dotty cosmology" as "no more relevant to real cosmology than the computer game *Life** is to evolutionary biology."

UNDERSTOOD THE UNIVERSE, BUT FAILED AT THE TEA POT?

Understandably, Disney has his doubts about the simulations of the whole universe, since the physics of one single star is hard enough to comprehend. Could we, for instance, simulate the physical processes in the sun on a supercomputer? It's not that easy. The problem is that normal matter does nastily complicated things—nuclear reactions, light radiation that depends on the wavelength and on the opacity, heat conduction, turbulent flows, supernova explosions creating heavy elements, which then in turn influence heat conduction in the next generation of stars…and so on. And dark matter? Nothing but gravitation. That's it. No wonder that the calculations become conveniently simple.

We should not lose sight of the fact that computers, as wonderful as they are, do have their limitations. Let's consider a humble example. Could we calculate the boiling temperature of water using the characteristics of a H_2O molecule? To our surprise, we can't. This problem is too difficult for the machine.

Vaporizing water, for instance, is a complex, virtually chaotic behavior of matter, as is the freezing when the molecules once again show their fractal structures (e.g., snowflakes!). This unpredictable process seems to happen during the formation of galaxies, too. It is therefore rather naïve to entrust the understanding of the universe's evolution to the computer.

But, credulity aside, can we at least trust that the observations justify the models? Unfortunately, these work best where there is barely any data. Incidentally, the most successful simulations of astrophysics take place in the so-called dark age. This is the period between the era of the cosmic microwave background, 380,000 years after the Big Bang (with a redshift of $z = 1,100$), and the emergence of the earliest visible galaxies half a *billion* years later (at $z = 10$). Whatever you may simulate there, telescopes won't protest

JOHANN JAKOB BALMER: A BRIGHT MIND IN DARK TIMES

To understand the dark age—when the hot plasma era ended and the universe turned transparent and dark—you have to know a bit about the

---

* An extraordinary simplistic cellular automaton invented by John Conway.

behavior of hydrogen atoms. We return to the story of Johann Balmer, a Swiss math teacher who in 1885 investigated light emitted by hydrogen. When light rays pass through a glass prism and hit a screen, the same wavelengths gather on one line, and for that reason are called spectral lines. Hydrogen alone, though being the simplest atom, shows an irritating multitude of spectral lines. The phenomenon seemed so complicated that spectroscopy was regarded as an imprecise, wishy-washy field among physicists. But Balmer was driven by a unique curiosity. It seems almost incredible how he was able to discover that the wavelengths were related to differences of the reciprocals of squares ($\frac{1}{4}$, $\frac{1}{9}$, $\frac{1}{16}$). Try to find this out without a calculator! It was a tremendous accomplishment, surely accompanied by great intuition.

Niels Bohr, by another stroke of genius, explained this when he developed the quantized atomic model in 1913, based on Balmer's formula. Electrons in atoms like to sit at the lowest possible energy, much like water seeks out the lowest spot. When excited electrons drop down to discrete lower energies, they emit light at a frequency directly related to the distance between allowed energies. And when they happen to absorb light of a suitable frequency, they jump up again! In Balmer's honor, the spectral lines originating from the jump down to the first excited state (called $n = 2$ in the notion of atomic shells) were named the Balmer series.

Others followed this pioneering achievement, including Theodore Lyman, after whom the series arising from electron jumps to the ground state $n = 1$ is named. Note that Nature is providing us with a very specific wavelength that predominates in the whole universe, the one that is generated by the simplest excitation of the simplest atom. It is called the Lyman-alpha line and is located in the ultraviolet wavelength at about 122 nanometers.

A GLIMPSE OF THE EARLY DAYS

With help of the Lyman-alpha line, we'll now take a deep look into the universe. Cosmologists are interested in the dark age between the redshifts $z = 1,100$ (start) and $z = 10$ (end). In the era of the cosmic microwave background, positively charged protons and negative electrons merged into hydrogen atoms. Only thanks to the resulting electrical neutrality did the universe become transparent. Light coming from the first quasars (the "background lamps") should thereafter spread unhindered, except for the Lyman-alpha line. Its photons literally get stuck in every hydrogen cloud along the way to us.

However, the cosmos is expanding, and the Lyman-alpha lines are therefore significantly stretched, depending on the red-shift of the distant quasar. All its light has to pass through hydrogen clouds in our line of sight, but much of it gets absorbed by the clouds, which themselves reside at a certain redshift. In summary, every wavelength between the 122 nanometer line and the quasar light at, for instance, 850 nanometers has a hydrogen cloud at a suitable distance, which absorbs the light the quasar had emitted. Thus, in principle, almost all wavelengths should disappear! We would see nothing.

Luckily, there are gaps in the hydrogen cloud distribution arising from clustering at later times. As a consequence, there isn't a light absorber for every redshift any more, and eventually, we see certain wavelengths. Put briefly, the quasar spectra called "Lyman-alpha forest" show that the universe has gotten holey and contains vast empty spaces. It has turned into Swiss cheese.

THE VOIDS' VOID UNDERSTANDING

We know that gravitation causes matter to agglomerate, but unfortunately no one really understands why the voids that result from agglomeration are so utterly vacant. It is as if a cosmic vacuum cleaner has been at work. Not even the computer simulations can reproduce this. Did gravitation work differently in the early universe after all?

These models of structure formation, resting upon the standard picture of the cosmos, fail to explain this emptiness. Hydrogen should still be present in these voids, although significantly thinned, but enough for absorbing the quasar light. But we still see the quasar light, unhindered by hydrogen! So what now?

To escape from this dilemma, people considered hydrogen present in an ionized form, for then it would be no hindrance to quasar light again. But how could re-ionization—the separation of negatively charged electrons from the positively charged protons—have taken place, and when? Well, possibly in the dark ages following the plasma era, and maybe through radiation of the first stars? But to imagine that almost all atomic partnerships between protons and electrons in the entire universe were somehow again split up in this dark age surely needs getting used to.

Methodologically, it is dangerous to quickly postulate a complicated mechanism because one *cannot* see something. And, of course, it is simply comfortable to put the mechanism in a hard-to-observe epoch. The voids could simply be empty, because we haven't understood the whole process of structure formation.

DON'T UNDERSTAND IT? GET THE COMPUTER!

Let's have another look at the kind of odd ideas produced when the silicon brains take over. For so much re-ionization of hydrogen to happen, it is supposed that a certain type of old stars, called population III, existed in the early universe. (Relatively old stars in globular clusters belong to population II, while the sun, living in the disc of a spiral galaxy, is happy to be a member of the young population I.) But due to the lack of data, astronomers prefer to simulate the Methuselah stars of population III in computer models.

Recently, a researcher claimed to have found a Population 2.5 (without Roman numerals) using his computer. This is not a satire on numerical results, but an article in *Nature*. Dan Hooper, one of the young stars of the dark ages and author of the book *Dark Cosmos,* recently came up with an even more interesting idea: since we don't understand the origin of re-ionization, maybe it could have been caused by stars formed out of dark matter.[5] Dark matter to dark stars during the dark age. Well, why not? The German philosopher Georg Christoph Lichtenberg once said that looking in a dark room for a dark cat, which isn't even there, is metaphysics. Bit by bit we are moving toward that.

SPECULATING WITH DARK CAPITAL

Though all this modeling seems to have gotten a little out of hand, physicists who don't embrace the dark outgrowths of the standard model are viewed with suspicion. Martín López Corredoira, an astronomer from the Institute of Astrophysics of the Canary Islands, wrote an article that gave a telling account of the irritating situation in *The Sociology of Modern Cosmology.*[6] His book *The Twilight of Scientific Age* should worry everyone about the state of affairs in today's science. There is often little time or interest in alternative theories, especially ones that would undermine the considerable time and effort expended on the standard model of cosmology.

It is not surprising that the first hints of a fractal structure of galaxy distribution, which means contradicting the dogma of the homogeneity of the universe, evoked fierce reactions. Francesco Sylos Labini's supervisor, Luciano Pietronero from Rome, presented his findings at a conference in Princeton in 1996. But he really shocked the experts by dissecting in detail all the hastily advanced counterarguments. Nonetheless, his idea of fractal structures is still far off the mainstream. Even João Magueijo, himself an "outlaw" cosmologist, wrote, "If this is true, start crying convulsively."[7] I once confronted Francesco: "Don't be surprised that everyone is mad with you," I said. "After all, your fractal universe would crash the standard model and would therefore pull the rug out from under every mainstream cosmologist. Why don't you at least suggest an alternative model?"

Francesco had a very dry answer:

> Listen, I am not the snooty-nosed making up theories. I am just doing a couple of statistical calculations I have learned to do. I don't understand much of cosmological models. And do you really believe that alternatives would have a chance? These people are used to fiddling around with their 95 percent dark substances so haphazardly that they can always describe any data they want. Why the hell should I bother?

In a way, I can understand him. He feels like an honest craftsman whose business underwent a hostile takeover by some debt-funded holding. We are buying dark matter and dark energy on credit, but it isn't clear at all if it will ever be cashed experimentally. Besides, everybody is doing it.

IT'S UNITY THAT MAKES THE WEAK'S POWER
(FRIEDRICH SCHILLER)

The tragedy of this collective cosmology conducted on the "we'll pay the bill later" plan is that, apart from a few exceptions, scientists are hardly foolish sheep. Far from it. In fact, their work distinguishes them from the many ridiculous pastimes of today's civilization. Most researchers are trying to properly conduct their work with the best scientific intentions. But barely anyone can claim to being completely unaffected by the opinions in their scientific environment. This has nothing to do with dishonesty, but is a sociological fact.

Would it really be appreciated if a young researcher conducting a project on "Numerical simulations of sypersymmetry-dark energy-couplings and their signatures in the Lyman-alpha forest"* paused to wonder out loud if dark matter even exists? Rather, you will find papers in which, for instance, a group of 47 authors (16 of which do not fit on the title page any more) "proves" the particle nature of dark matter through an extensive simulation of gravitational lensing.[8] Being in a choir makes it easier to be a believer, and being in a large choir makes it pretty much mandatory.

> Immaturity is the incapacity to use one's intelligence without the guidance of another.
>
> —*Immanuel Kant, German philosopher*

---

* This is a title I made up, but not without a basis.

A BRILLIANT DESCRIPTION OF THE EARLY EON, OR
JUST LIPSTICK ON A PIG?

I apologize for such a highly opinionated attack, but it does appear to me that the pendulum has swung much too far the other way.

—*Mike Disney*

The standard "concordance" model of the universe serves thousands of scientists with bread and work, and the hope of that work being reflected upon is correspondingly tiny. The most skilled will implement the latest techniques for their computer simulations, and continue tinkering with the theories, which should be scrutinized instead. A quote attributed to Einstein says, "Any intelligent fool can make things bigger, more complex, and more violent. It takes a touch of genius—and a lot of courage—to move in the opposite direction." The odds of moving in a new direction are long, since numerical astrophysics is booming. The science critic John Horgan may be overstating the case in his book *The End of Science* when he writes, "Computers may, if anything, hasten the end of empiric science."[9] But unfortunately, there is a grain of truth.

Of course, computers are a wonderful tool for doing science today. I am enthusiastic about a computer algebra system that, for instance, helped me analyze gravimeter data. But one of the big issues with simulations is that we can't be sure if the underlying equations of gravity are correct. And computers do everything but point this out. They do what they are told to do. As Pablo Picasso said, "Computers are useless. They can only give answers." There are plenty of indications that gravitation needs to be corrected at small accelerations. As long as we don't understand this, no computer power whatsoever will suffice. All the really nice animations, computer-generated images, and cleverly applied new parameters may turn out to be nothing more than a pig in a poke.

Chapter 10

SPECULATION BUBBLES RISE

EXPANSION, IMAGINATION, INFLATION: DO WE KNOW THERE WAS A FIRST SECOND?

For a long time, cosmology struggled with an important question. If the expansion of the universe is slowed down by the gravitational attraction between everything in the universe that has mass, is there enough mass to halt the expansion or even to reverse it? This scenario of a "closed" universe, in which mass will act as a brake and stop expansion, is contrasted with an "open" one, in which the expansion is too fast to be ever halted by gravitational attraction. The point at which the universe is poised between being open or closed is called a "flat universe." This state, which requires a particular mass density, is extremely unstable, on par with a needle balancing on its tip. Just as the needle will fall at the slightest disturbance, the teetering universe would quickly have to determine its fate of being ever-expanding and "open," or being "closed," which means it would be bound to crash.

Cosmologists try to measure the average density of mass in order to see which alternative is going to be realized. This yields an irritating result. Although the data slightly favors an open universe, this "decision" between open and closed should have been made far earlier by the cosmos. It is as if we are now observing a needle leaning toward one side after balancing on its point for the last 14 billion years. It was in 1969 when an extraordinary physicist, Robert Dicke, pointed out this riddle of the incredibly balanced universe. Dicke had achieved great accomplishments in experimental physics, but his deep thoughts about gravitation are underestimated to this day. His discovery is referred to as the "flatness" problem, because the case at the edge between the expanding and the contracting universe is described in space and time by the theory of general relativity with a vanishing curvature.

However, at the same time the theory can't explain why the universe should be in such a delicately poised but highly unstable state.

SMOKE SIGNALS COMING FROM ALL SIDES.
WHO GAVE THE COMMAND?

There are more mysteries of the universe to come. Basically, the cosmos appears the same from whatever direction you may contemplate it, which is called "isotropy." But two of the farthest-away galaxies, each on opposite sides of our cosmic panorama, cannot know of each other. There isn't enough time since the Big Bang for light (information) to travel that distance,* and thus nothing that happens in the one galaxy could influence the one on the other side. This effect is still more dramatic in the cosmic microwave background, which tells us something about the early universe. Here, regions just two angular degrees apart cannot know anything about each other. Why then does the cosmic microwave background appear to us as a perfectly blended soup everywhere in the universe? How did the universe manage to agree on an overall temperature of 2.73 degrees Kelvin, today's observation? This puzzle is called the "horizon problem."

Although the universe's "flatness" is to be understood figuratively, it may be visualized by referring to the curvature of a sphere's surface. In case of a large sphere such as Earth, we perceive its local surface to be flat rather than curved, even though we know it is curved (except for adherents of the Flat Earth Society). Now the grandiloquent, though quite banal, idea to fix the flatness and horizon problem is to postulate that the universe—analogous to a sphere or a round balloon—suddenly inflated in an early period, and that everything that was close together and uniform was suddenly far apart and uniform. Such an "inflation" would, however, cause a growth of approximately 30–50 powers of ten at about 10^{-35} seconds after the Big Bang. Besides such a beginning being rather exotic, the contradictory observations in the early universe we have discussed in the previous chapters should make us reflect upon the whole model.

It would be quite reasonable to infer that perhaps the theory of general relativity cannot be fully correct. However, because this would be too much of an earthquake in science, people would rather cook up an ad hoc story

---

* Technically, it is a more subtle task to compute the travel time from the distance, but we don't need to bother with this here.

such as inflation. Going to such early times is a huge extrapolation, and theorists can find comfort in putting this inflation into a period shortly after the Big Bang, where nobody will ever observe it.

INVASION OF THE UNIFIED THEORIES: THE FOOLHARDY CONQUER ASTROPHYSICS

The MIT physicist Alan Guth, the creator of inflation theory, is a high-energy physicist. Traditionally, in this field, people don't distress themselves too much when making daring extrapolations toward GUTs and TOEs (Grand Unified Theories and Theories of Everything). While scrambling for such an ambitious goal, many theorists back in the early 1980s let go of any pretense of modesty.

According to Guth, "These spectacularly bold theories attempt to extend our understanding of particle physics to energies of about 10^{14} gigaelectron volts."[1] This may be spectacularly higher than the energy of one gigaelectron volt, which corresponds to a proton's mass. But is the hope of a mouse swimming across the Atlantic truly spectacular? Or just stupid? Anyway, it's easier to make a start as a group. So Guth teamed up, as he recounts, with a small drove of physicists who began to dabble in studies of the early universe.[2] "The goal of the cosmological research involving grand unified theories," said the American physicist, "is to solidify our understanding back to 10^{-35} seconds after the Big Bang."[3]

It is precisely solid knowledge that is in short supply here, because replacing observations by theorizing means giving up the scientific method. But boldness has surely grown beyond the greatest theories. Let's listen to Guth again: "Physics was now actually ready to talk about these bizarre sounding events in the Universe,* fractions of a second and even billionths and billionths and billionths of a second after the Big Bang. Absolutely fantastic."[4]

Of course, it's fine to talk about one's fantasies. For science this holds true, with tiny restrictions, such as what we may actually observe. Perhaps Guth should have listened to the philosopher Ludwig Wittgenstein, who said, "Whereof one cannot speak, thereof one must be silent."

REFLECTION: IS ZERO A GOOD VALUE TO BE MEASURED?

Roger Penrose, in his brilliant book *The Road to Reality*, dug into Guth's inflation theory and picked it to pieces with a series of arguments. To begin

---

* The pattern in the cosmic microwave background is often referred to as sound waves.

with, argues Penrose, inflation is only possible by introducing a scalar field (which assigns a number to every point in space), "unrelated to other known fields of physics and with very specific properties designed only for the purpose of making inflation work."[5]

There is actually a more severe problem. Does it make sense to talk about the curvature of the universe and then state, parenthetically, that this curvature has been measured to be zero? (This results in a "flat" universe.) People act as if this zero is just a number like the 43 arc seconds of Mercury's perihelion advance, or like the mysterious fine structure constant 137.036 in atomic physics. However, mathematically, the number zero indicates a very special case, and treating it like any other number means that too little reflection has been done.

If you repeatedly determine the volumes of geometric bodies to be zero, you would, after a while, also suspect that someone might have foisted two-dimensional planes upon you instead of true three-dimensional objects. What we would really need in order to understand the universe's zero curvature is a theory that excludes curvature by construction. The theory of general relativity instead allows curvature. The deficiency of our comprehension of gravity may lie right here. Instead of doubting general relativity, people assume that the universe inflated. Why not steamrolled flat by a track paver? Puffing the universe with inflation is just a patch—accompanied by a lot of puffery. Rather than resorting to such fancy, we should question the naïve belief in a Big Bang independent of the flow of time.

WE ARE BRITISH

On close inspection, it is hard to explain the homogeneity of the cosmic microwave background, even with inflation. "Whatever this generic singular structure is," said Penrose, "it is not something we can expect to become ironed out simply because of a physics that allows inflationary processes."[6] He considers the entire idea that the puzzling uniformity resulted from a heat transfer (the blended cosmic soup) a flawed concept. What a pain in the neck it must be for a leading mind of thermodynamics like Penrose to see the growing popularity of inflation! Nonetheless, being the British gentleman that he is, Penrose spends five sentences apologizing before carefully stating, "Since I believe that there are powerful reasons for doubting the very basis of inflationary cosmology, I should not refrain from presenting these reasons to the reader."[7] In plain words: he thinks the whole notion is garbage.

As the British cosmologist Martin Rees recounted in his book *Our Universe and Others*, Penrose was more outspoken when talking about inflation: "Inflation is a fashion high-energy physicists visited on cosmology. Even aardvarks think their offspring are beautiful."

Rees himself, *Astronomer Royal* and Fellow of the *Royal Society*, was gentlemanly as well in expressing his reservations. "All these ideas highlight the link between the cosmos and the microworld—the ideas won't be firmed up until we have a proper understanding of space and time, the 'bedrock' of the physical world."[8]

In general, European physicists prefer to word the theses of inflation in conjunctive mood, peppered with lots of "woulds," "coulds," or "shoulds," a habit that is usually washed off on the way over the Atlantic. This may be the reason why a much cheekier critic of inflation, João Magueijo, wrote:

> As years went by, inflation's popularity among physicists continued to grow. Eventually, inflation itself became the establishment. So much so that it gradually became the only socially acceptable way to do cosmology—attempts to circumvent it usually being dismissed as cranky and deranged. But not on the shores of Her Majesty, Queen Elizabeth II.[9]

DON'T BELIEVE IN UNICORNS? INFLATION HELPS

We will get back to Magueijo later on. At this point, I can no longer resist showing you the most beautiful argument in favor of inflation. It was the prime reason why inflation theory has had such appeal for particle physicists.

A magnet has both north and south poles, but if you cut it in half, you will never be left with isolated poles. You'll just have two smaller magnets with their own north and south poles. The reason behind this is that magnetic fields are caused by moving charges; according to Maxwell's equations of electrodynamics, no single magnetic charge or monopole can exist on its own. Looking at it formally, Maxwell's equations would be more symmetric if there were single magnetic charges analogous to the electric ones. Symmetry is considered beautiful and fundamental, and therefore physicists have been hunting for monopoles for a long time with their high-energy colliders. Despite intense efforts, experimenters have come up empty-handed. At this point, perhaps we should just take Nature as it is and accept that it may have compelling reasons for why it doesn't entirely lump together the electrical with the magnetic field.

Theorists are not always good sports. Even if the experimental evidence is against their ideas, as clearly as it is, they do not blame their neat theories but rather accuse Nature of not behaving properly. We may assume, for instance, that monopoles do exist, but their density in the universe has gotten so low through the enormous expansion during inflation that we have little chance to observe them today. And now the argument's jewel in the crown: the fact that we *cannot* see monopoles is strong supporting evidence for the theory of inflation!

For those of you who are not yet suspicious of proof through nonobservation, recall that with the same justification, you could claim that the early universe was full of unicorns, and it is solely by the (un)fortunate event of inflation we don't see them now. You may shake your head, but this is a common type of conflation in modern physics. You can become a theorist and invent any hokum new particle you want, and if there isn't any experiment capable of detecting it, inflation will surely come to the rescue. No doubt inflation is so popular in particle physics for that very reason.

INFLATION HIGHEST BIDDING—UNIVERSITIES ARE STANDING IN LINE

The history of inflation theory provides some insights about the methods and structure of high-energy physics in the late 1970s. When Alan Guth told his colleague Leonard Susskind his idea for the first time, Susskind wryly commented, "You know, the most astonishing fact is that we are even being paid for such things." I'm not sure why Susskind lacked these healthy intuitions in his later days. And indeed, he may have changed his opinion soon afterward.

The physics community reacted quickly to Guth's inflation proposal. He was immediately offered jobs at every renowned American university, including Harvard, Princeton, Stanford, Cornell, the Fermilab, and the Universities of Pennsylvania, Minnesota, Maryland, Columbia, and California at Santa Barbara. He ended up with a professorship at MIT the day after he applied. When you're hot, you're hot!

But remember, this was all for a highly speculative idea that would never ever have the chance of being tested experimentally. We must conclude that the way fundamental physics was practiced at the leading institutions was already sick back then.

Guth's book *The Inflationary Universe* gives you an impression of the turmoil of calculations he grappled with for weeks. Today, it seems as if a theoretical physicist is all the more esteemed the more complicated and impenetrable his calculations become. We need to be careful not to mistake complexity for profundity, as the philosopher Karl Popper once noted.

Not contradicting yourself is already considered a success. Whether the results have anything to do with reality is considered a minor issue. Apparently, this thinking had become so widespread that we can hardly blame Guth for priding himself with his discovery. The particle physics Nobelists Sheldon Glashow and Murray Gell-Man promptly congratulated Guth on his inflation solution. Steven Weinberg, another Nobel laureate, was even jealous that he didn't think of the idea himself. Maybe this sudden love of cosmology emerged because they were weary of their models of particle physics back then.

MODERN HERETICS AND POSTHUMOUS CENSORSHIP FOR EINSTEIN

In the 1990s, Great Britain was one of the last refuges where inflation was still considered as American hype, and thus João Magueijo was able to deal with an alternative idea, a theory with a variable speed of light. The idea is quite interesting, although Magueijo failed to point out earlier approaches taken in this direction, such as by Robert Dicke[10] and even by Einstein.[11] Magueijo's proposal is still green, and more than the theory itself, one should admire that he managed to publish it in *Physical Review*, a decidedly conventional journal.[12] You might compare that to putting an ad for sex toys in a Catholic Sunday paper.

The speed of light is, by the way, not a measurable value in today's physical unit system. Rather it is defined at 299,792,458 meters per second, while the units meter and second are together determined by the oscillations of a cesium atom. Were these time and length scales to change slightly, we would not even notice a change of the speed of light in meters per second because meters and seconds are plainly changed, too.

Nonetheless, the idea of variability in the speed of light continues to be a red rag for most theorists. One South African scientist, in a comment on Magueijo's proposal, claimed that such an idea would undermine the theory of special relativity.[13] Interesting! Are we to understand here that Einstein, who had a try on a variable speed of light in 1911 as well, didn't have a clue about the basics of his own theory?

THEORY OF INFLATION TURNS INTO THE INFLATION OF THEORIES

The first theory of "inflation" proposed by Guth quickly turned out to be contradictory, and that was immediately repaired by some even more exotic assumptions. Since then we are experiencing an orchidaceous growth of chaotic, eternal, and other variations of inflation theory. Needless to say, the theory generously helped itself with free parameters, in order to at least fit the data to a certain degree. This enables theorists to explain observations in retrospect, if they had been incompatible with the theory beforehand. The cosmologist John Barrow wrote, "This elasticity has diminished the faith of the general astronomical community in inflation, and even led some researchers to question whether inflationary cosmology is a branch of science at all."[14] Recently, Paul Steinhardt, one of the founding fathers of inflation theory, gave the answer in an article: inflation is not falsifiable.[15] (Steinhardt's newly favored theory is even weirder, by the way.)

But just as with astrology, the theory seems to live off of the lousy memory of humans. In an article in 1989, Guth said that the spatial frequencies of the microwave background (the typical spot size in the pattern) must have the same strength at every scale, which is in blatant contradiction to the WMAP data.[16] In contrast, the fine-tuned flatness of the cosmos confirmed by this data was publicized as evidence for inflation, as if the flatness problem known since 1969 was something new. In the near future, it will surely be announced that some of the results of the Planck mission had been known beforehand by inflation.

AND THE LORD SPOKE: 0.96

Who wants to betray people has first to make the absurd seem reasonable.
—*Johann Wolfgang von Goethe*

When scrutinizing the predictions of inflation regarding the cosmic microwave background, it turns out that they are merely a clever way of saying nothing. If you don't know anything, then simply forecast complete randomness. Usually, this is packed in the phrasing that the data is "consistent" with the predictions of inflation. Sounds good, but it merely states that inflation is not disprovable at the moment and probably never will be.*

The data of the cosmic background radiation, even disregarding the uncertainties mentioned in chapter 8, show that some sizes of patterns are more prominent in the sky than others. To express it mathematically, a so-called scalar spectral index amounts to 0.96 instead of 1, which would mean the sizes have equal strength. But of course, the various subspecies of inflation theory have their parameters at hand, and with a large helping of mathematical jiggery-pokery, they can show that the "predicted" spectral index coincides with the current measured value 0.96. Should we thus believe the number 0.96 to be a logical necessity, a consequence of the notion of an early-on, superfast expanding universe? What kind of fairy-tale story are we being told here?

---

* If this seems to be a positive statement to you, you had better be careful! Most of the nonsense in the world is not disprovable. We will come back to this issue in chapter 16.

Fig. 10.   A well-known visualization of the universe's evolution. A lecturer at the St. Petersburg conference in 2008 compared it to a tipped-over beer glass. As one easily recognizes, the cosmic microwave background is at the bottom. Someone from the audience interjected that inflation theory must then be the science of more than one glass of beer—or vodka.

The mathematical gimmickry brushes aside the fact that the data of the microwave background originated 380,000 years after the Big Bang. Hence, the imagined time period for inflation of 10^{-36} seconds until 10^{-33} seconds after the Big Bang is almost 50 decimal powers earlier than the curtain that hides the early universe from us. Go figure. To claim that anyone can deduce a traceable assertion out of this within reasonable error margins is plainly ridiculous.

Slowly but steadily, inflation has morphed from an exotic attachment to the standard model of cosmology to an accepted part of it. As Bertrand Russell expressed so beautifully, "The fact that an opinion has been widely held is no evidence whatever that it is not utterly absurd." Inflation relentlessly gains ground, to the extent that reasonable people such as Lee Smolin have, in the meantime, come to regard it as "experimentally well secured." This leaves us with the picture of a spreading disease, a type of collective Alzheimer's of theoretical cosmology. The healthy ones like Penrose belong to the soon-to-be extinct species.

Chapter 11

BLACKING OUT

BLACK HOLES, THE BIG BANG, AND QUANTUM GRAVITY: ECOLOGICAL NICHES FOR THEORISTS

It is an irony of fate that the term "Big Bang" was coined in a radio show in 1950 by the astrophysicist Fred Hoyle, who, in his time, was one of the harshest critics of the notion that the universe began with that sudden expansion. Hoyle wanted to mock the model that was, in his eyes, far too simple.

What does "Big Bang" actually mean? Despite all the unsolved mysteries of how the structure of galaxies and clusters formed, we can undoubtedly see an expansion of the universe. The cosmic microwave background hardly allows any conclusion other than that the universe used to be smaller, hotter, denser, and more homogeneous. All these observations are fed by the model of the Big Bang, but it is also evident that we know less and less the further in time we go back to a hotter and denser world.

We reach the dividing line between science and speculation when we get to densities that are higher than that of an atomic nucleus. There are simply no observations in that realm. Yet inflation theory has exceeded this limit by many orders of magnitude and extrapolated backward down to the Planck length of about 10^{-35} meters.

The Planck length, by the way, doesn't have anything to do with the early universe. Rather it follows from combining the fundamental constants G, h, and c in a way that yields a unit of length, the meter.* The utter impossibility

---

* The square root of $Gh/2\pi c^3$ has the unit of meters.

of measuring anything at the scale of the Planck length, which is 10^{20} times smaller than that size of a proton, has not deterred scientists from speculating extensively.

Thus the Planck scale has become a particularly popular playground for prestigious institutes such as the Department of Applied Mathematics and Theoretical Physics (DAMPT) in Cambridge (UK), the Perimeter Institute in Canada, and the Institute of Advanced Studies at Princeton. It is ensured that no experimental result will ever stain their prestige.

After all, there are serious doubts whether the Planck length has any fundamental significance, as the Nobelist Sheldon Glasgow has pointed out.[1] It is merely a limit beyond which we do not know anything. A hallmark of excellence in theoretical physics.

A BRIEFER HISTORY OF QUANTUM GRAVITY

Many hammer all over the wall and believe that with each blow they hit the nail on the head.

—Johann Wolfgang von Goethe

Since the Planck length contains the gravitational constant G and Planck's quantum h, it is the scale at which "quantum effects of gravity" are supposed to become important. Dear readers, that is all. No theory of quantum gravity exists, let alone any evidence of an observable effect. You'll find the topic somewhat more elaborated in Stephen Hawking's book *A Briefer History of Time*. The chapter about quantum gravity takes up 21 pages, of which almost 20 pages are devoted to repeating gravitation and quantum theory. Andrzej Staruszkiewicz, the editor of a renowned physics journal, commented on this topic:

"It is tempting to assume that when so many people write about 'quantum gravity,' they must know what they are writing about. Nevertheless, everyone will agree that there is not a single physical phenomenon whose explanation would call for 'quantum gravity.'"[2]

There is simply no theory that combines general relativity with quantum theory. All theoretical recipes cooked up until now have failed, as for instance the so-called ADM formalism, a reformulation of Einstein's equations of general relativity. It has gotten nowhere, but among theorists, it is nevertheless considered a bible leading the way. Another great couturier of theories is Abhay Ashtekar, a widely cited physicist who regularly praises the accomplishments of the Loop Quantum Gravity he fathered. His summary: "It is interesting."[3] Actually, it has been interesting for 30 years now. Not exactly news anymore.

Of course, some theoretical problems with quantum gravity have been tackled brilliantly since then, but they are brilliantly irrelevant when it comes to understanding gravity. Quantum gravity theorists deserve credit for their justified criticism of string theory, yet their own field doesn't have any graspable results. Success just comes in different flavors.

DOES THE GRAVITATIONAL CONSTANT CEMENT THE FAILURE OF QUANTUM GRAVITY?

Problems cannot be solved by using the methods which created them.

—*Albert Einstein*

Right from the start, theoretical physicists seem to have deployed a number of assumptions that block deeper reflection. For instance, there is the belief in an unalterable gravitational constant G. Solely for this reason, all theoretical attempts trudge through the Planck length's eye of a needle and come across as trying to push a door that is labeled "pull."

Maybe quantum effects do not start at what Roger Penrose calls the "absurdly tiny" 10^{-35} meters, but at the nuclear scale of 10^{-15} meters. I can hear the outcry from theorists, but by no means do we understand this realm yet. Nuclear physics gratifies us with hundreds of measured values that describe the excitations of the nuclei, but none of them can actually be calculated.

While Niels Bohr's quantum theory marvelously derives energy levels for the atomic shell out of the constants of nature, nuclear physics has not achieved anything comparable yet. I don't want to attach blame here, but one must soberly admit that there is no theory making predictions from first principles. As long as such a theory doesn't exist, someday gravity may well become significant for nuclear physics.

EXPERTS OF KNOWLEDGE AND EXPERTS OF PUBLISHING

The fact that no theory for quantum gravity exists does not preclude the existence of numerous quantum gravity experts. For those who have encountered such experts, I want to specify. The science historian Federico di Trocchio distinguishes experts of the first and second category. The first are specialists who are familiar with specific problems, such as professors of geodesy, hydraulic engineering, or even quantum optics. In contrast, according to di Trocchio, there are also experts of the second category, whose knowledge will become immediately obsolete once the riddles scientists are studying have been understood. They make their living on the problems that are being tackled unsuccessfully. Usually this manifests itself in many publications, indicating "active" research fields. Hence, there are experts on dark energy hypotheses, experts on quantum cosmology, and, of course, experts on the so-called Theory of Everything. Although this ultimate goal of physics has been announced by many, including Stephen Hawking, I fear that we will have a long wait. So let us rather look back at a real expert of quantum theory.

ON THE CLIFF ABOVE THE OCEAN: THE BIRTH OF QUANTUM THEORY

In the spring of 1925, Werner Heisenberg was tormented by a severe attack of hay fever. He fled to Helgoland, a small German island in the North Sea, where the absence of pollen should have helped him recover. His nose was so violently swollen that his landlady suspected him of having been in a student brawl. Completely undisturbed, Heisenberg pored over the unresolved riddles of atoms. After working on his calculations until sunrise one night, he found out how to describe the puzzling quantum behavior of matter, though the arising formulas were awkward.

Few physicists had a strong mathematical background in the 1920s, and while unwittingly having reinvented matrix algebra to explain electron orbits, Heisenberg had little confidence in the muddle he had come up with. Fortunately, his supervisor, Max Born, was a trained mathematician who immediately recognized the mathematics involved and was able to formalize Heisenberg's model (a fact that helped math to be taken seriously by physicists).

Bohr's intriguing model of the atom was hereby vindicated, and even Einstein, as stubbornly skeptical toward the new theory as he was, recognized Heisenberg's spectacular success: "He has laid a big quantum egg." Einstein himself, with his famous formula for the energy of light quanta, $E = hf$ (where f is the frequency), had discovered the fundamental importance of the constant h.* But Heisenberg, in 1927, elicited an even more profound meaning from it: h represents the limit of what we can know with certainty—the Heisenberg uncertainty principle.

The quantum h is the barrier of the microscopic world that ultimately hinders our knowledge. If the position of a particle is known, its momentum (the product of its mass and velocity) is not precisely measurable—as

---

* Einstein was awarded the Nobel Prize in 1921 for this photoelectric effect, because the committee was unable to make sense of the theory of general relativity.

is the case when an electron passes through a narrow slit. Metaphorically speaking, every particle resists being captured by jiggling around, a consequence of its wave nature.

The same limitation applies to the relationship between time and energy, two quantities that relate to cosmological conundrums, as mentioned in chapter 4. Following Heisenberg's principle, you can borrow an amount of energy E out of nothing, if it is readily paid back after the time $t = h/E$. The uncertainty principle is one of the most important laws of quantum theory, if not all of physics.

QUANTUM OF SOLACE: HOW TO ESCAPE FROM BLACK HOLES

When Stephen Hawking reflected on black holes in the 1970s, Heisenberg's uncertainty principle came to mind. It says that the energy borrowed out of the vacuum of space can produce particle pairs when energy is converted into mass using the equation $E = mc^2$. Normally these particles would immediately merge again and disappear suddenly, and therefore remain inconspicuous. However, in a strong gravitational field near a black hole, it may happen that one particle partner bolts into it, while the other manages to make an escape. But particles cannot originate out of nothing, so the unpaid energy bill has to be taken over by the black hole. Thus, Hawking concluded, there may be a net escape of particles from black holes through this quantum effect, which is forbidden by the classical laws of gravity.

As neat as this thought might be, it is a far cry from every observation, for a black hole with a solar mass would then need 10^{66} years to evaporate by ejecting particles. At least Hawking's theory is useful to calm people who were afraid of particle accelerators producing mini black holes, because they would decay very quickly according to Hawking's equations. Nevertheless mini black holes belong to exotic theory constructs that have never been observed. As usual, such things are assumed to have existed at the Big Bang.

THEORIES TRAVELING AT THE SPEED OF THOUGHT

Even an excellent law of Nature such as the uncertainty principle can be perverted by implementation at the Planck length. A single particle confined to a distance of 10^{-35} meters would have the hazardous energy of a Formula One racing car. This is surely somewhat remote from the experiment and a good thing at that, too. But why not combine the Planck scale with the Big Bang? Squeezing all particles of the universe into 10^{-35} meters, instead of solely one, is even more absurd, but, being a theory, it is hazard-free.

A physics research group at the Max Planck Institute in Munich developed a model "bridge to the Big Bang" that resolved—at least that's what we read—the unification of gravity and quantum theory. Their website modestly states: "Scientists explain with string theory how the universe developed right after its birth." The cosmos used to be shriveled due to its initial minor size because of quantum fluctuations, but was then smoothed out over time. Wow! A long time span corresponds to smaller energy fluctuations, an outright rediscovery of the Heisenberg uncertainty principle. Unfortunately, the Nobel was already awarded in 1932 for this, but the embarrassing nonsense found its way into a media release, and, even more disconcertingly, into *Physical Review Letters*.[4]

> There are people who catch fish, and people who just muddy the water.
>
> —*Chinese proverb*

A LITTLE STEP FOR A MATHEMATICIAN, A GIANT LEAP FOR A PHYSICIST

The road to the Big Bang is long and weary. We have to go back in time to the dark ages, cross the opaque curtain of the cosmic microwave background, and then immerse ourselves in the hot plasma soup. Then we get to a confusing variety of effects for which our knowledge is blurring into vague speculations, such as the aforementioned "shriveled cosmos" era.

Mathematicians are diving into this muddy water because their own interests are coming to life again at the beginning of time, when $t = 0$. At that point, the density of the universe is infinite, and time is not defined anymore. Physics loses its meaning here. However, a mathematician will gloat over the problems of the Big Bang "singularity," the mathematical jargon for infinite values.

Physicists indulgently mock mathematicians for the joy they find in their calculations. They tend to think of mathematics as merely providing a toolbox of techniques for solving physics problems, much to the irritation of mathematicians, who are inclined to view physicists more as workaday mechanics making practical use of the delightful array of sophisticated ideas they have developed.

I am a partisan, of course, but this largely amicable love-hate relationship is illustrated by many jokes at the expense of the other side.

A balloonist who is lost in the fog exclaims, "Where am I?" and gets the response, "You are in the basket of a hot-air balloon!" This statement immediately identifies the respondent as a mathematician; it was thought out, correct, and absolutely useless for the solution of the problem.

To come back to Werner Heisenberg here, one has to consider the beginning of his university career a stroke of fortune. At the recommendation of his father, he visited the famous mathematician Ferdinand Lindemann. Lindemann was an eminent scholar who had solved a century-old problem concerning π, but had a rather condescending attitude toward a greenhorn such as Heisenberg, with his ambitious physical ideas. The conversation was full of misunderstandings, and finally the nonstop yapping of Lindemann's lap dog compelled Heisenberg to go. Frustrated, he changed his mind and found his way to physics. Some say that this was Lindemann's greatest achievement.

BLACK HOLES: THE ASYLUM FOR CALCULATIONS

Mathematicians love the Big Bang's singularity, and they just cannot resist racking their brains about it and about black hole singularities. The "radius" of a black hole* is important for computing the effects of general relativity, but it is fruitless to discuss physics inside that radius. We cannot learn anything about the interior of black holes because no information can escape its gravitational prison. End of discussion...except for all the ensuing discussion.

Since their discovery, black holes have attracted an interest somewhat disproportional to the evidence for their existence. Their size has never actually been measured. And interestingly, black holes with a significantly higher density than that of a nucleus haven't yet been observed.[5] I can see the theorists cocking their eyebrows, but observations simply end here.

But back to the experiments. Technologically, we are far from realizing the conditions inside a star in a controlled way. Otherwise, nuclear fusion plants would be fueling the energy needs of the planet. Thus, it is quite presumptuous for us to believe that the Big Bang itself is being "simulated" by the current highest-performance experiment of particle physics, the Large Hadron Collider (LHC) at CERN. And it certainly motivates us to have a close look at another boldly media-amplified result, the Higgs boson, which we will scrutinize in the next chapters.

Let us take a moment to review the fantastic expectations raised by the LHC that we have seen vanish into thin air. Michio Kaku, a famous theoretical physicist, predicted in a television interview that the LHC would prove string theory, allow us to look at the time before the Big Bang, discover whether our universe has collided with others, and finally figure out what happened before Genesis 1:1.[6] Any questions? Kaku will give an answer,

---

* The black hole radius is called Schwarzschild radius $r_s = 2GM/c^2$; see chapter 5.

no doubt. But even people such as Stephen Hawking or Lisa Randall have made numerous promises about what the LHC is going to discover. Sadly, the only real thing being observed is an industry of theoreticians transforming the unexpected into the predicted with the benefit of hindsight.

> Physicists are thus led to try safe experiments which are obvious and acquire their interest from the unusual energy region at which they are performed, being otherwise rather simple-minded.[7]
>
> —*Emilio Segrè, Nobel laureate, statement in 1980*

DOES THE LHC TUNNEL LEAD TO A PARALLEL WORLD?

Media attention is an important thing. Rolf Landau, the head of the CERN public relations division, boasted about the universe being contained in the volume of a pinhead. Despite this being a fairy-tale story, it doesn't have anything to do with CERN. It is therefore wise to leave still more exotic speculations to people who are less in the limelight. In the LHC video "Big Bang in the Tunnel," cute PhD students with winning smiles (but no publications yet) are talking about their LHC dreams. One scene particularly delighted me. The student suggested that one could test a parallel world of some inflation theory. "There might be a world where Napoleon *won* the battle of Waterloo…and another one in which America is still British territory." Yes, that would be great! And maybe in this other universe, such silliness wouldn't have infiltrated physics.

Whatever one may want to say against CERN, it is an experiment. If you are benevolent, you may excuse the grandiose claim of simulating the Big Bang as necessary propaganda in our era of overhyped media. At least the ever-increasing energies that are part of the LHC experiment are real. My first-year professor in electrodynamics phrased it waggishly: "Here we have come a little closer to the question about what the world holds together in its inmost folds! From 10^{-16} to 10^{-17} meters…"

It does, however, become rather preposterous when progress is taking place on a mental level only. A prospective assistant professor at the University of Pennsylvania working in this fashion was described as follows: "He managed to come closer to the Big Bang as than it has ever been achieved, through *his equations*." This reminds me of my son, who, as a three-year-old, undertook journeys by putting his finger on a map, settling on Australia, and saying, "Daddy, did you know that I was with the kangaloos?"

I can, right now, get us even closer to the Big Bang by following my son's approach. The magic is to start at the equations $t=5$, $t=4$, $t=3$, and keep on going!

BEYOND TOTAL IGNORANCE: THE NEW EINSTEINS

Regardless of our mathematical progress, we should be ready to meet absolute ignorance at the Big Bang. So what skills should you bring along to explore the time before the Big Bang? A negative IQ? Or at least a purely imaginary one? Stephen Hawking, who is not entirely hostile to speculation, writes that such guesswork about the time before the Big Bang has no place in a scientific model of our universe. But maybe he just wasn't aware of the book* *Once Before Time: A Whole Story of the Universe*. Its author, on the cover of his book, is referred to as "Einstein's successor."** That's basically the whole story of the book.

John Baez, also an advocate of Loop Quantum Gravity, maintains a checklist called the "crackpot index," in which he ranks the muddle-headedness of wannabe scientists. He scores publications with the following: 3 points for a logically inconsistent statement; 20 points for every use of science fiction works or myths as if they were fact; and 30 points for suggesting one's work is similar to Einstein's. The theories of shortly after, during, or before the Big Bang accumulate quite high scores.

---

*To be historically fair, we must concede that even this nonsense pioneered with the string theoretical book *The Universe before the Big Bang*.

** German original cover. And there is another successor we shall encounter in chapter 15. What an era we are living in!

Chapter 12

THE FIANCÉE YOU WON'T MARRY

THE STANDARD MODEL OF PARTICLE PHYSICS: HOW PLAYING WITH MATHEMATICAL BEAUTY TOOK OVER REAL LIFE

"Gentlemen, a university isn't a bathhouse!" David Hilbert, one of the most ingenious mathematicians, was upset. His famous phrase was the last desperate attempt to get the university administration in Göttingen to award a faculty position to his assistant, Emmy Noether, who had some brilliant accomplishments behind her. Female professors, however, were still unimaginable in the conservative atmosphere of the German Empire in the early twentieth century.

Hilbert had good reasons for his commitment. Noether, who had to continue to lecture as an assistant, discovered a fundamental theorem about symmetries in physics, which became one of the most influential ideas of the past century. What are symmetries? They are easy enough to understand if you recognize that the laws of physics do not depend on, say, where you set the zero-point of the x, y and z axes in three-dimensional space. It is fine anywhere, whether it is the middle of the Sun, at Polaris, or in your kitchen. You are free to choose the place you want to start your measuring. Symmetry is when something does not matter. It turns out that for the laws of physics, neither position nor rotation matters.

Emmy Noether deduced that, as a consequence of the latter rotational symmetry, there is a quantity that is *conserved* in mechanical motion called angular momentum. This is the product of mass m, velocity v, and the distance to the rotation axis r. It is the law that allows figure skaters to accomplish such dizzying spins. They stretch their arms out at the start of the spin, then pull them in to their bodies (the axis of rotation). Their mass can't change, so as r gets suddenly smaller, v gets sharply faster. Angular momentum is conserved under rotation. The law of the conservation of energy, still more importantly, follows from nothing more than that the laws of Nature are symmetric in time and thus stay always the same (although this is surely not self-evident in cosmology).

You might get an impression that theorists love symmetries! They do. When searching for new laws of Nature, they single out equations with intrinsic symmetries in space and time, and this paradigm has been predominant in physics for about 50 years. Though this is done everywhere by everybody in physics, let us see if it is wise.

THE WHOLE IS MORE THAN THE SUM OF ITS PARTS

I think I can safely say that nobody understands quantum mechanics.

—*Richard Feynman*

Before talking about the evolution of symmetry concepts, let me briefly tell you about an experiment that contains the heart of quantum theory. It is called the double-slit experiment. Since it made theoreticians fall into despair in the 1920s, every physicist knows it, though no one can claim to deeply understand it.

In the macroscopic world, if you use an air gun to shoot at a plate with two narrow slits in it, there are simply two zones behind the plate where it is better not to stay. If you place a screen behind the plate, you will see two belts hit by the bullets. Now go to the microscopic world and do the same thing with a laser pointer and very tiny slits less than a micron wide. What you see on the screen will be entirely different! Rather than showing two stripes, the laser light will produce a peculiar pattern that proves the wave nature of light. While the wave nature of light was known for a long time, the truly sensational result of the double-slit experiment is that a wave pattern appeared when scientists fired *electrons* through the slits! Elementary particles behaved as if they were like light. This phenomenon has become known as "wave-particle duality." Revealing the true nature of electrons was worth a Nobel Prize in 1929 for Louis-Victor de Broglie, after whom the wavelength of elementary particles is named.

Mathematical physicists like to see a symmetry embedded in this wave nature of electrons. Waves have troughs and crests, but all the experimental results do not depend on that—only the overall strength of the wave (amplitude) is important, while its phase (the position of the troughs and crests) does not matter. Now make a virtue of necessity. If the wave phase doesn't matter, it's a symmetry! This is puzzling, though in a way it's just another aspect of the puzzle that particles sometimes behave as waves.

SPIN: SOMETHING ROTATING, BUT NOT LIKE A TOY

As if the wave characteristics of particles weren't strange enough, elementary particles act even more eccentrically. For instance, electrons have what's called "spin," which you may imagine as a rotation on its own axis. But this is not the whole truth. If you want to know the direction of the rotation axis by putting them close to a magnet, you will see that they line up either parallel or antiparallel to a magnetic field, but never at an angle to it.

This stubborn behavior of electrons, discovered in 1922, created a major headache for physicists because it highlighted that something was wrong with the picture that electrons simply rotate. It seems that electrons (and other elementary particles) need a double full turn, that is, 720 degrees of rotation, to find themselves in the original position, whereas all everyday objects will show you the same face again once you have turned them by 360 degrees.

Hadn't Emmy Noether's beautiful theorem stated that the laws of Nature were related to a rotational symmetry in space? Now, the electrons forced mathematical physicists to invoke an additional symmetry to account for the particles' desire to turn twice. As a consequence of this crazy way double rotation, the electron carries only *one half* of Nature's fundamental angular momentum, which is again provided by Max Planck's constant h.* This appears to be something really basic. But despite all the sophisticated math describing it, nobody understands why Nature invented all of this.

HOW A GENIUS MAY FOOL HIMSELF

By the way, Einstein made an interesting mistake while measuring the electron spin in 1915. Together with the Dutch physicist W. J. de Haas, he designed an experiment in which they tried to turn electrons in a metal upside down by using magnetic fields. On the one hand, the electrons had

---

* Precisely, the angular momentum is h divided by 2π. For simplicity, we will just say h here.

an angular momentum, and their inner rotation generated a magnetic field analogous to orbital currents. Einstein and de Haas wanted to determine the ratio of the magnetic field and the angular momentum and were expecting it to be 1, because they had not yet discovered the spin. They were still confident of playing with little toy tops. In one measurement run they obtained a value of 1.02, and in another 1.45. They quickly convinced themselves that the 1.45 value was an apparatus error and published the ratio 1.02, allegedly complying with the theoretical value, which is not 1, but 2, as found out later.

Ultimately, this mistake did not do much harm to physics. But it highlights that one must be careful when a result agrees with the expectations of the experimenter. Also keep in mind that the setup was rather simple here, Einstein was a conscientious physicist who had little difficulty with challenging mainstream thinking, and he certainly didn't need to seek out fame, because he was already famous. We should not take for granted that these three things automatically apply to today's situation in astrophysics and particle physics.

SYMMETRIES ALL OVER THE PLACE: WHERE IS THIS JOURNEY TAKING US?

Let us now look at how particle physics got tangled up in too abstract concepts. The idea of symmetries in physics started to take on a life of its own when scientists began to examine the interior of atoms more closely. Consider the example of neutrons, which are nuclear constituents like the proton, but electrically neutral instead of positively charged.

Unfortunately, a single neutron has a half life of just about 10 minutes, and then it decays into a proton, an electron, and an antineutrino—a process known as "beta decay." The decay process, not fitting into the common scheme of a force, is called "weak interaction." Why it occurs on average after 10 minutes but not, say, after 20 minutes is unknown. Even the very reason why neutrons don't live forever is a mystery. The only thing we know is that during the beta decay process, there is a neutron or proton in play, before or after—it doesn't matter. Now catch a breath. This is a symmetry.

But this is also where physics went astray. If we are honest, invoking such a symmetry (which was called "isospin") explains almost nothing. But it allowed physicists to relate the abstract symmetry of the beta decay with the symmetry of the electron waves. Based on this analogy, Sheldon Glashow, Abdus Salam, and Steven Weinberg developed a model that claims to combine the electromagnetic force and weak interaction. They won a Nobel Prize in 1979 for this. All efforts at finding unifying theories in physics nowadays almost exclusively resort to these avenues using abstract symmetries, administered by a field of mathematics called "group theory."

In particular, a lot of new particles have been classified using mathematical symmetries. There is, however, a methodological pitfall. The abstract math does not explain the measureable quantities of particles.

Let's look at a comment made by Howard Georgi, a physicist who had worked in the Grand Unified Theories (GUT) business inspired by group theory, and therefore not the most radical science critic: "Symmetry is a tool that should be used to determine the underlying dynamics, which must in turn explain the success (or failure) of the symmetry arguments. Group theory is a useful technique, but it is no substitute for physics."[1] Dan Hooper, having mutated into an astronomer, varied Galileo's words that the book of Nature was written in math language. "Group theory," he said, "is the dialect of particle physics." Wolfgang Pauli, the Nobel Prize winner of 1945, gave a somewhat harsher diagnosis. He called the upcoming fashion "group pest."

There is only one step from the sublime to the ridiculous.

—Napoleon Bonaparte

THE GRANDIOSE UNIFORMITY...OF THOUGHTS

Nevertheless, it is by now established wisdom that the electromagnetic and weak interactions are unified through the group theoretical description of their symmetries. But even the far looser connection to the nuclear force is regarded a "unified" model today, merely because one can formally stick together their mathematical symmetry groups. Richard Feynman cheekily wrote about this abstract reasoning of his Nobel Prize colleagues: "If you just look at the results they get, you can see the glue, so to speak."[2]

The euphoria over the invention of still more general and abstract symmetry groups came to dominate theoretical physics. It culminated with the 2004 Nobel Prize. The official announcement stated that the honored researchers "have brought physics one step closer to fulfilling a grand dream, to formulate a unified theory comprising gravity as well—a theory for everything." Obviously, Stockholm was dreaming, too...

QUANTUM ELECTRODYNAMICS: THE DANCE OF ELECTRONS AND LIGHT

It should not go unmentioned here that theoretical physics also had achieved great accomplishments. Long before the symmetry fashion took over, Richard Feynman became famous for his intriguing interpretation of the interactions of electrons, positrons, and light.

The basic idea is fairly easy to grasp. Thanks to Heisenberg's uncertainty principle, a traveling electron can borrow for a little time t an amount of energy $E = h/t$. Electrons may use this energy for juggling with photons. Like two people sitting on wheeled office chairs who are throwing heavy medicine balls to one another and rolling backward every time they pitch or catch the ball, two electrons that exchange photons knock each other back, too. Feynman managed to reformulate the laws of electrodynamics— two electrons feel a repulsive force—in these funny terms.

The calculations based on this have led to predictions that have been precisely tested and are considered the best-measured results of all physics.* Richard Feynman, Julian Schwinger, and Sin-Itiro Tomonaga were justifiably awarded the Nobel Prize for this in 1965. The big insight of the theory is that light and the most basic particles, electrons and positrons, show such a puzzling similarity. Yet nobody knows the reason for it.

THE COLORFUL ORNAMENTATION OF QUANTUM ELECTRODYNAMICS

What wonders is the manner in which this theory succeeded for six decades to keep our understanding within the limits set by quantum field theory.

—Anthony Leggett, Nobel laureate

In former times, classical physics upheld the picture that it was the electric and gravitational fields in space that caused the accelerations of charged particles. Quantum electrodynamics completely abandons this idea in favor of the exchange of borrowed photons. Feynman's theory worked so well that particle physicists decided to use it as a blueprint for all other interactions. Though the old wave-particle quantum theory of Bohr, Heisenberg, and Schrödinger should be a caveat against describing everything with particles, the idea entered the back door and seized hold of modern physics.

But unlike quantum electrodynamics, the results of its extension to nuclear physics, called quantum chromodynamics, are anything but precise.[3] It is therefore utter speculation that imposing the concept of quantum electrodynamics on atomic nuclei is the right way to go. Nevertheless, theorists almost exclusively walk on this well-trodden path.[4]

---

*The magnetic moment of an electron (its inherent magnetism) and the so-called Lamb shift in the spectral lines of a hydrogen atom.

Though the experimental agreement is disappointing, usually the "uniformity" is praised as a flash of inspiration. Very funny is a comment of Feynman on how his own ideas were pushed to a too general level:

So when some fool physicist gives a lecture at UCLA [University of California Los Angeles] in 1983 and says, "This is the way it works, and look how wonderfully similar the theories are," it's not because Nature **is** really similar; it's because the physicists have only been able to think of the same damn thing, over and over again.[5]

The 1960s saw nuclear and particle physics in a quite desperate situation. Particle physicists sought to reveal the interior of atomic nuclei by smashing them up using particle accelerators, but the collisions were revealing hundreds of short-lived elementary particles. Who could make sense of such a bunch of very poorly understood objects? The science writer James Gleick wrote:

[As scientists] pushed more energetically inside the atom, they were watching the breakup of the prewar particle picture. With each new particle, the dream of a manageable number of building blocks faded. In this continually subdividing world, what was truly elementary?[6]

Murray Gell-Mann and George Zweig, who would later receive the Nobel Prize, tried to clean up the mess and introduced hypothetical quarks to explain the composition of these many new particles. To fit the experimental data, quarks would have to carry noninteger multiples of the elementary charge e. In this case, the proton, with a positive charge of $1e$, would be made up of two *up* quarks with a positive charge of $2/3e$ each and one *down* quark with a negative charge of $-1/3e$. To hold the quarks together, so-called gluons were introduced to the theory. They take the same medicine-ball role in the nucleus as the light quanta did for the interaction of electrons. According to quantum chromodynamics, quarks are the ultimate building bricks of heavy matter.

Using various combinations of quarks, physicists managed to classify most of the newly discovered particles, but nevertheless the idea was still widely regarded as a theoretical chimera. Things changed in 1974, when a group of researchers at the Brookhaven Laboratory announced the discovery of a new type of quark and antiquark pair. Later, experimentalists found a third pair. The particle zoo continued to grow.

It was always the best intentions which led to the worst creations.

—*Karl Kraus, Austrian poet*

SIMPLER, BUT NOT REALLY SIMPLE

As Murray Gell-Mann frankly admitted during a talk in Munich in 2008, Heisenberg considered the entire idea of fractional charges assigned to quarks to be nonsense. Had the genius Heisenberg already become a senile, stubborn skeptic at the age of 65? It is unlikely that he felt biases against fractional quantities, as he, in his freshman years, had proposed the famous "half-integer spin" on an electron, which back then stood in sharp contrast to the established wisdom. However, half-integer spins make sense observationally, whereas no one has ever seen a fraction of a charge.*

This is perhaps the most absurd shortfall of the model: isolated quarks don't exist. Quarks always appear as pairs or triplets, and there is no way to break them apart into singles. No one knows why, although a neat term was invented to describe it: confinement. Does it make sense to talk about parts of something that cannot be divided into parts? Another very nasty problem arose because the quark model had to care about an exclusion principle in quantum theory found by Wolfgang Pauli. It forbids identical particles living together in the same place. To get around that problem, physicists had to invent another messy attribute in order to make it okay for identical quarks to live together in the same place. They invoked *colors* so that quarks living together would no longer be identical, but as Roger Penrose relentlessly reminds us, they are "in an essential way unobservable."[7] The colors, in brief, were introduced solely to keep the theory from contradicting itself. And once again we face an example where contradictions were removed by inflating the model.

In the canon of particle physics, mostly told in retrospect, each step seems reasonable, but it is sobering to follow the detailed account in the book *Constructing Quarks: A Sociological History of Particle Physics* by Andrew Pickering. Pickering is a highly qualified particle physicist, but his experiences in the field led him to investigate the mechanisms for how today's mainstream concepts were established. He demonstrates the way successive adaption of techniques that are sensitive to the desired effect greatly influences what is generally thought to be an experimental fact. The countless erroneous interpretations, contradictions, ad hoc assumptions, and widespread herd mentality described by Pickering raise serious questions about the entire method of high-energy physics.

---

* There is, however, an interesting and quite embarrassing story to tell here, outlined in an article by Andrew Pickering (*Isis* 72 [1986]: 216–236).

A THEORY OF ANYTHING

If we try to summarize the standard model of particle physics, we count six quarks, each in three different "colors" (red, green, and blue), along with their antiparticles. This adds up to 36 kinds of heavy particles, not a small number. Let's not forget the gluons gluing the quarks, which have to be colorfully dressed (e.g., red and antigreen). Imagine for a moment you are hearing this for the first time. Does this sound like Nature speaking to you?

For the greatest minds of physics, simplicity was the supreme law of Nature, but during decades of busy research in particle physics, that law gradually faded into oblivion. It's not that physicists are stupid and don't realize at all that there might be something wrong with such a complicated and arbitrary construction like the standard model. But they tend to repress the fact that if the standard model is wrong, it has to be discarded as a whole, and this prospect is too scary. Thus even if no one is really convinced that it is "the" correct theory, the standard model, being the perpetual fiancée of physicists, offers a secure living to many. One comes to an arrangement, but no one searches for explanations for the multitude of concepts, although this would be the real mission of a theoretical physicist. "Truth, if ever, is found in simplicity, never in the multiplicity and confusion of things," Isaac Newton said. He has never been as right as today.

Chapter 13

CHRONICLE OF A SURPRISE FORETOLD

HOW HIGGSTERIA DELAYED THE BANKRUPTCY OF PARTICLE PHYSICS

It seems to me that particle physics is the one area with the greatest unsolved intellectual problems. Some of them are almost a century old, and physicists tend to forget them because there is no inkling of their solution.

—*Emilio Segrè, discoverer of the antiproton, in 1980*[1]

The standard model of particle physics is unable to predict the observed masses of its particles. This is really quite embarrassing, given that mass is such a basic property of particles. At the very least, the standard model needs a way of explaining mass, because without it, the universe would be a whirl of massless (inertia-free) particles buzzing around in space. OK, there is one idea. In 1964, the Scottish physicist Peter Higgs and others had proposed a particle that somehow imbues particles with mass necessary for anything in the universe, such as stars, Earth, and us, to exist. For that reason, the Higgs particle is sometimes called the "god particle."*

To get an impression of how the Higgs idea works, let's go to Oktoberfest in Munich, the largest folk festival in the world. You are in midst of a lot of people. It's a very popular event and it's not easy to walk through the crowd.

---

* Leon Lederman, the author of the (very good) book *The God Particle*, had originally proposed the title "the goddamn particle."

The crowd represents the Higgs field. It restricts your movements, which means you have inertia, or equivalently, mass. Now imagine Madonna walking through the Oktoberfest crowd. The people gathering around her would dramatically slow down her possible motions and thereby increase her mass. It is her celebrity status in the Higgs field that acts as a mass. The Higgs particles themselves would correspond to propagating little crowds of people retelling and discussing how Madonna was dressed.

But how would you go about finding a Higgs particle? Physicists at the Large Hadron Collider (LHC) at CERN have been slamming protons together at higher and higher energies with the hope of finding a trace of something that might hint at the existence of the Higgs. Keep in mind that the huge number of proton collisions produces a lot of "background" noise in the detectors. That noise is 12 orders of magnitude larger than the signal that would identify the Higgs. It is as if from the 7 million liters of beer drunk at the Oktoberfest, a single drop of an unexplained nature leaked out. That's what happened at CERN, but it sufficed to intoxicate particle physicists.

Data analysts would argue in all seriousness that they know how much beer was in the barrels, could measure how much accidently spilled on the floor, and so on. It is, mildly speaking, ambitious to claim that one could control such a tiny leakage.

> Simply the fact that scientists mobbed together in agreement is enough to warrant caution and modesty.
>
> —*Newspaper comment in 2012*

TOO MUCH FOREST TO SEE THE TREES

> Most particle physicists think they are doing science when they're really just cleaning up the mess after the party.
>
> —*Per Bak, Danish theoretical physicist*

Researchers have to hunt through a vast amount of data produced by the collisions, looking for signs of the Higgs. It seems that the enormous difficulty of analyzing the LHC data keeps the experts so busy that they have no time to read textbooks any more. As we have seen in chapter 4, electrodynamics still has some severe inconsistencies that do not allow us to calculate the precise amount of photons emitted when electrical charges are accelerated (or decelerated). Richard Feynman and Lev Landau agree on this fact,[2] but it has not sunk in with the particle physics community. Today's colliders generate the highest accelerations ever produced by humankind. How can one predict the number of emitted photons in an experimentally unprecedented situation, if there is no valid formula at hand? It's just ridiculous.

There is another very odd thing in the Higgs observation, one that is not often noted amid all the celebrations. The excess of photons in the experiments doesn't produce a nice peak on a graph at a certain energy, but a broad hump. This not only suggests quite a large uncertainty for the mass of the sought-after particle but also indicates a very short lifetime of around 10^{-25} seconds, about 1,000 times shorter than predicted by the standard model. This is a simple consequence of Heisenberg's uncertainty principle.

Particle physicists are arguing that the hump-shaped signal is due to the yet-insufficient energy resolution of the detectors. (As a matter of fact, in 1974 in Brookhaven, the mass resolution was 20 times better.[3]) Thus, to match the predictions, the signal peak must still become 1,000 times narrower. It is surely not an easy task to bring it down to this level, thus high energy physicists prefer to talk the problem down.

While in this case one can resort to the claim that the Higgs may have a longer lifetime than it is visible from the data, it is generally acknowledged that another particle of the standard model, the top quark, does have a lifetime in the range of 10^{-25} seconds. Such a particle can at best travel the distance of the diameter of a proton and never get into a detector.* Thus, all the "evidence" is indirect inference depending on the correct modeling of hundreds of millions of protons crashing into each other every second. And remember, we don't even know the radius of the proton precisely![4] It's absurd to believe one can exclude all possible artifacts that could easily arise in the complicated analysis. Such systematic errors are grossly underestimated with respect to statistical ones (that are proudly presented to be minuscule). This fallacy of neglecting the unknown unknowns with respect to the known, easily calculable unknowns is a main topic in Nassim Taleb's book *The Black Swan*. High-energy physicists seem to be completely unaware of the risks they are taking. Their entire method of data analysis has become unreliable, a fragile construction.

VARIABLE REALITY, CONSTANT EPIPHANY

What is there to celebrate, though? In some ways, the most honest answer is: we're not quite sure yet....But there are ample grounds for jubilation, whatever the new particle turns out to be.

—New Scientist, *July 2012*

Let's take a look at the nature of the signal being sought in the collider experiments. It consists of two extra photons. As a nonphysicist you may

---

* Considering relativistic time dilatation doesn't change much.

be surprised, but almost every particle-antiparticle pair decays into two photons. So what was the big discovery?

The excess in the two-photon or "diphoton channel" data where the photon pairs are measured was initially twice as high as predicted by the standard model (such discrepancies tend to melt away when refining the model).

The standard model had predicted production rates* for other pairs, such as electron-positron, but some of these rates just did not match the observations. What happened then? People were, bizarrely, interpreting the corresponding "underproduction" or "overproduction" as "tantalizing hints for new physics." Wow! This means congratulating ourselves that we still don't understand what's going on. Any discrepancy in the data can be fudged in this way.

What is the probability of being successful in such a poorly defined search? I believe it is high. What we are observing in the aftermath of the 2012 Higgsteria is a meticulous analysis of mostly irrelevant details, such as testing if the Higgs has "spin zero" (when it was quite clear beforehand that it did), and an overselling of every banality by statements such as "we have come closer to the detection," or "The particle has become more Higgs-like" and so on, as if this would mean anything.

High energy physicists are a lot like art enthusiasts desperately seeking to discover a van Gogh painting. Once they find a painted canvas they claim there is strong evidence that it could be a van Gogh. The next steps are announcing "Oh, yes, we proved it's painted in oil," or "we found blue color on it—very typical for van Gogh" and then rushing to a press conference with the news that "all our findings so far are in perfect agreement that this is a van Gogh." To be fair to the art enthusiasts, the difference here is that *there are* some real van Gogh paintings.

"Establishing" the discovery means reaching the point in which everyone becomes tired of hearing such announcements. And everyone is happy once the painting is finally sold as a van Gogh to some uninformed buyer—in the Higgs case, alas, it's the public who pays.

Is it really very hard to see that this is merely a sociological process, a development of language rather than of physics? The actual convention of speech, first used by CERN director Rolf-Dieter Heuer, is that the signal is "a Higgs particle," not necessarily "the standard model Higgs particle."

But hold on. The discovery of a non-standard-model Higgs is still a triumph of the standard model, right? Or is it the triumph of a nonstandard or substandard standard model yet to be developed? The bitter truth is that

---

* So-called branching ratios.

theoreticians will instantly invent new patches to modify that vague thing as soon as it becomes necessary. The standard model can actually accommodate *every* result, like an adaptive virus in the scientific method, resistant to any falsification. There is no way to doctor that conclusion.

THE WHATEVER PARTICLE—AND A MASSIVE BLUFF

The Higgs doesn't take us any closer to a unified theory than climbing a tree would take me to the Moon.

—*John Horgan, science writer*

I hope I haven't annoyed you with too many technicalities, but the real flaw of the Higgs particle is that the whole idea is banal. Even if we suppose that the LHC "discovery" had made sense, it is a great self-delusion to think that it would help to understand what mass is. No one has been able to explain the mass ratio of proton and electron, which is about 1836. If we beg to know, at long last, why the proton is so much heavier than the electron, we would get the enlightening answer that it is because the proton couples 1836 times stronger to the Higgs field.* It's better for physics not to have that kind of "progress" in understanding.

Sadly, this embarrassment is not just the fault of the Higgs particle, but a defect of construction of the standard model. It is completely unable to predict the masses of the elementary particles it describes, and to make it even worse, the model does not even provide a vague idea of how to deal with the problem. To give an example, muons and tauons are particles very similar to an electron, but for some unknown reason much heavier. Richard Feynman wryly commented, "Nature gives us such wonderful puzzles! Why does She repeat the electron at 206 times and 3640 times its mass?"[5] Among today's great minds, it seems that nobody cares.

Lee Smolin describes a scene that apparently took place in the office of John Preskill, a leading particle physicist. In 1983, Preskill jotted down the mass problem on a little sheet of paper for the fun of it, and stuck it to his bulletin board in order to keep the essential problem in mind. Fifteen years later, the note was still in its place but faded beyond recognition!

A couple of years ago, rumors were going around about a gigantic computer simulation of quantum chromodynamics, which promised to calculate the masses of the neutron and proton, along with their ratio to the electron mass. What would be truly revolutionary is if the respective mass ratios of 1838.68 to 1836.15 could be determined *theoretically*!

---

*Some would say the quarks couple, but this does not really make a difference.

The grandiloquent announcement of the simulation results, published in *Science* in 2008,[6] turned out to be not much to boast about. The calculations were only accurate to 4 percent, which is rather poor when you consider that you need an accuracy of 0.2 percent to even distinguish between a proton and neutron. So far, I haven't heard of any group rushing toward a testable result. And regrettably, as everywhere in such simulations, nobody can control what the computer code really did.

AT LEAST PROVE YOUR INABILITY

The standard model, because it excludes gravity, is an incomplete account of reality; it is like a theory of human nature that excludes sex.

—*John Horgan, science writer*

There is a very general argument over whether computer calculations for mass can work. All a computer can do is spit out numbers, but the physical unit of a mass (kilograms) has to be generated by a combination of fundamental constants. There is no way around it, unless you cheat. If you take the constants of nature—speed of light c, charge of the electron e, Planck's constant h, and the permittivity of space ε_0, there is simply no combination whatsoever that yields the unit of mass. You can try it out for yourself.

The only way to arrive at a unit of mass would be to include the gravitational constant. But first you have to find a unified theory containing gravitation and quantum theory, and we are still a long way from that happening. Many physicists dodge the problem by claiming that physics does not permit the computation of mass. But this is just a way of confessing intellectual impotence. The job of theoreticians is to explain Nature, not to find excuses why they can't.

For a solution, one probably has to contemplate the units of the fundamental constants, but some people maintain that calculating mass is precluded in principle, like squaring the circle, a math problem the world has been hunting after for centuries in vain. If this should be the case in physics, please demonstrate the impossibility! The mathematicians have done their homework here and proved it using a characteristic of π.*

Given that the standard model barely predicts anything, it is grotesque how many physicists kick around and claim how "precisely tested" it is. But if you ask, you never get an answer expressed with specific numbers, which would actually mean something. The precise tests are just an example of something that became true by endless repetition in parrot fashion.

---

* That was the great achievement of Ferdinand Lindemann, the professor, whose dog had the merit of having guided Heisenberg's talents to the right faculty.

FORBIDDEN MASSES AND THE UNBEARABLE
LIGHTNESS OF BEING

We haven't talked about the neutrino much, a constituent of the standard model that again struggles with its mass. It is another add-on to the main structure of the standard model. Three types of neutrinos have long been established and—though by quite indirect evidence—they seem to transform into each other. Physicists have theorized that a neutrino must have a small rest mass by arguing that the mechanism of transformation does not allow for massless particles. At the same time, the original standard model prohibited neutrinos from having rest masses. But since the theory already has so many bits pinned, stapled, and glued to it, one more modification didn't really bother anyone.*

The number of free parameters in the standard model of particle physics—almost 20, and far more if you count the masses—has long reached an unsettling magnitude. The lion's share of this complication is due to the quark model, even if publicized as a simplification. Some simplification is not too hard to achieve, if one had accumulated heaps of poorly understood data as particle physics did in the 1960s. So is such excessive gathering of material then really helpful for science? David Lindley, the former editor of *Nature* and *Science* and one of the few reflective voices about the Big Science business, said, "Untrammeled experimental discovery is not an entirely good thing: the array of newfound particles overwhelmed attempts to produce economical theories."[7]

Actually, real progress in history was most often achieved when comprehension wasn't lagging too far behind observations. Evidently, too much undigested information creates the danger of mistaking banal descriptions for theories. The parallel to the epicycles of Ptolemy, that overcomplicated model of medieval astronomy, is obvious, but the cautionary example of the deadlocked geocentric worldview with its many parameters isn't realized in the busy high energy physics community. What happened in both cases was that people confused the model with the physical entity it meant to describe.**

Particle physics seems to be a dead loss, since people have become used to not scrutinizing questionable concepts such as "colors" or "confinement" any more once they have been accepted. Erwin Schrödinger, one of the founders of quantum mechanics, had severe difficulties with the far

---

* There is a result, however, that, if verified, would cause a lot of trouble for the standard model: the so-called neutrinoless beta decay.

** The British philosopher Alfred North Whitehead called this the fallacy of misplaced concreteness.

less obscure image of electrons suddenly changing their atomic orbits, an early example of a poorly understood concept: "If these damn quantum jumps don't come to an end, I regret having ever bothered with quantum physics!"

COLLECTING INSTEAD OF THINKING: THEORIES ON THEIR WAY TO THE PACK RAT

Maybe professional ethics such as Schrödinger's would benefit today's theorists. Ernest Rutherford, who had discovered the nucleus' size, once said that science is either physics or collecting postage stamps. Ironically, he got the Nobel Prize not in the field he esteemed most, but in chemistry. However, "collecting" certainly does not have a negative meaning anymore. In 2008, the Nobel Prize was awarded for the CKM matrix (a description of how quarks change their "flavors"), which contains some of the notorious free parameters. It is not really an essential building block of Nature. At least in 2012, the Nobel committee resisted the public Higgsteria and calls by people such as Stephen Hawking to award the Nobel Prize in October for whatever had been discovered at CERN four months earlier. However, particle physics lobbyists will continue knocking at the door and, having consumed the major part of funding for fundamental physics, also demand their share of the glory. There was too much money spent for not being rewarded.

The standard model of particle physics irritatingly reminds us of the confusing complexity of the financial markets. And unfortunately, even if thousands of people publicly agree on how well the market works, we have learned that it is not always wise to lean on that. According to Nassim Taleb, the more expert the people, the more prone they are to the epistemic arrogance that prevents them from being able to consider the possibility that they are mistaken. After all, most of the experts involved in particle physics have invested too much of their life in it to be able to question the credibility of the model as a whole. It has become too big to fail.

Chapter 14

NEW DIMENSIONS IN NONSENSE

BRANES, MULTIVERSES, AND OTHER SUPERSICKNESSES: PHYSICS GOES NUTS

> History suggests, however, that if these superparticles don't turn up, there will be strenuous efforts to save supersymmetry by tinkering with it rather than deciding that the whole thing is a failure.
>
> —*David Lindley, science writer, in 1993*

I am not old enough to remember all the deaths the theory of super-symmetry (SUSY) has died in the history of accelerator experiments. Supersymmetry attempts to give a unified mathematical description of elementary particles by postulating a collection of mirror particles. The hypothetical twins are distinguished from the known ones by their super-heavy masses, and for that reason, remain undetected so far in all particle colliders. The first big gravestone was the DESY lab in Hamburg in 1978, where supersymmetry grossly failed to predict the experimental outcome. But a really agonizing year must have been 2012, when *Scientific American* published an article in their May issue asking "Is Supersymmetry Dead?" and not the slightest trace of supersymmetry could be found in the long-sought-after Large Hadron Collider data at CERN.

Peter Woit commented in his blog[1] *Not Even Wrong*, "The paradigm that dominated the subject for the past 30 years has collapsed in the face of experimental (non)evidence, threatening to take down the life's work of hundreds if not thousands of theorists." However, the really die-hard adherents argue that "not the whole parameter space" has been checked yet, comparing the search for supersymmetry to a game in which your friend assures you that a pea is hidden under one of the five cups put upside down on a table. You have turned over four cups. Now do you still trust that the pea will show up under the fifth? Of course.

Indeed, there might be a remote possibility that some poorly defined signal is declared to be a supersymmetric particle, as was the case with the Higgs "discovery." The real problem is that supersymmetry continues to add cups to the game. There is nothing that prevents supersymmetry from justifying itself with the explanation that the pea was invisibly small and may be revealed by a new, more powerful accelerator.*

The whole game is not science any more. Take this for an example. Over the range of two years, a group of theorists, while facing the absence of evidence, continuously fine-tuned their predictions in a series of papers. They postulated ever-increasing energies from 455 to 708 gigaelectronvolts for the speculative particles, labeling their brain spawn with names such as "Universe F-U2," "Aroma of Stops," and "The Sweet Fragrance."[2]

Peter Woit said, "It's rather easy to extrapolate to the future what these authors will be claiming the SUSY masses are, harder to extrapolate how they'll be describing the smell."** We all know what perfume was used for in the French emperor's castles in the seventeenth century.

> Classical theory has it that a bogus hypothesis will be rejected when it fails to predict "reality." But such a catastrophe can be deferred almost infinitely by the elaboration of secondary hypotheses to explain why not fitting the facts is not—after all—fatal to the theory.
>
> —*Bruce G. Charlton, science writer*

THE ORIGIN OF ZOMBIE SCIENCE[†]

The reason behind the supersymmetry hype is the ugly complication of the standard model. With its arbitrary concepts, it leaves an unpleasant aftertaste, and this is an inspiration to look out for a more uniform, more beautiful theory, just like bleak industrial suburbs foster a desire for palm-decorated paradises.

---

*There is a nice April Fool's Day joke, arXiv:hep-th/0503249. Indistinguishable from today's science.

** Woit does a great job in debunking the string and SUSY crap. Unfortunately, he has pretty mainstream opinions with respect to the standard model.

† A term coined by Bruce G. Charlton, editor of the journal *Medical Hypotheses.*

But the journey's start takes place under hair-raising preconditions. One assumes, for instance, that the strength of the forces at play in particle interactions in the standard model varies with the energy level. However, this is nothing but a fudge factor, which can be easily played with to make contradictory observations comply with the model. The variation of these so-called coupling constants of forces is then extrapolated over 11 orders of magnitude to energies 100 billion times higher than those used in today's accelerators. Although this is already crazy enough, it doesn't even work! If it had worked, the higher and higher energies that represent going back in time to the higher energies of the early universe would mean the forces would all merge at some point in that early universe. However, since such a unification is highly desired, physicists felt free to install further parameters and tweak the model to their desire. This is what justifies the theory. Or, to put it less kindly, the naïve expansion of the standard model that doesn't work is sturdy evidence for supersymmetry. Right?

SUBPRIME PARTICLES: JUST RELABEL THEM

I would like to show you the conceptual banality of supersymmetry, but we first have to make a trip to Dhaka, back then in India, in 1923. The physicist Satyendranath Bose was reflecting upon the statistics of elementary particles and arrived at the following idea. Imagine a box with identical marbles. If they are labeled with numbers, you will be able to identify them even after a thorough mixing. But this is not how elementary particles behave. Once their waves interfere you cannot distinguish one from another anymore experimentally. Individuality is erased in the microscopic world.

Einstein was impressed by this insight and made sure that Bose's work was immediately published, even though several science journals had rejected the outsider from India. Unfortunately, neither Einstein nor Bose got to witness how their visionary prediction of atomic behavior was vindicated with laser techniques at ultracold temperatures, resulting in a Nobel Prize in 2001.

Bose was instrumental in sorting out particles into two different categories, one of which—the boson—was named after him. The second category—the fermion—was named after the Nobelist Enrico Fermi. The difference between the two, in terms of this categorization, has to do with the "spin" assigned to them.* Basically, bosons tend to cohabitate harmoniously like in the Nobel-worthy effect, while fermions, having a rather

---

* While bosons have integer spin h, $2h$, and so on, fermions have the values $1/2\ h$, $3/2\ h$, and so on.

difficult character, prefer to live in a space reserved for singles. Why Nature invented such a peculiar distinction is, by the way, not known.

Instead of digging into deep questions, supersymmetry just postulates that the bosons and fermions may interchange their spin behavior, so that bosons have the spin reserved for fermions and vice versa. This is nothing but a shallow reshuffling of Nature's laws. But once we give something a name, everything is fine. Particle physicists just put an "s" in front of the supersymmetric partner (an electron becomes a selectron) or attach the suffix "ino" to the end. We end up with impressive personalities with names like red-antigreen gluino or blue anti-charm-squark. Look! A whole new batch of elementary particles!

HOW TO ESCAPE THE EXPERIMENT—A CONTINGENCY PLAN FOR THEORIES

Though this be madness, yet there is method in it.

—*William Shakespeare*

The most annoying fact about supersymmetry is that the thousands of physicists who promote it for the past 30 years haven't been able to generate a single prediction that could genuinely test, that means possibly disprove its existence. Furthermore, even the most modest versions of the theory have introduced about 100 extra parameters describing particles that nobody has ever seen.

All this does not affect the popularity of supersymmetry among group theorists, who were happy to play with all kinds of mathematical rotations in a five-dimensional space—which, by the way, led to a series of apocalypses, thankfully only theoretical ones. These calculations can be arbitrarily tampered with when there is a need to escape an experimental result. Richard Feynman wittily describes how to avoid such a disaster:

> Somebody makes up a theory: The proton is unstable. They make a calculation and find that there would be no protons in the universe any more! So they fiddle around with their numbers, putting a higher mass into the new particle, and after much effort they predict that the proton will decay at a rate slightly less than the last measured rate the proton has shown not to decay at. When a new experiment comes along and measures the proton more carefully, the theories adjust themselves to squeeze out from the pressure.[3]

Feynman's mockery of physicists may seem frivolous, but that's how physics is done today. If something doesn't work, it is always Nature's fault and not the theory's.

DEEP QUESTIONS OF NATURE, FADED ON THE WALL

Instead of scientists imposing their arcane calculations on Nature, physicists in the early twentieth century listened to Nature's deep questions. Here is one of them. By combining the constants h, c, ε_0 and e in a certain way, a mysterious number comes out, the numerical value being 137.0359.... You may see it as a blend of quantum theory (h), electrodynamics (ε_0 and e), and relativity (c), which results in a pure number called the fine structure constant. In contrast to other fundamental constants of nature, this value remains the same even if you use different units in lieu of kilogram, meter, and so forth. Nevertheless we are talking about a value measured in a laboratory! Richard Feynman commented:

> It's one of the greatest damn mysteries of physics: a magic number that comes to us with no understanding by man (...). We know what kind of a dance to do experimentally to measure this number very accurately, but we don't know what kind of dance to do on the computer to make this number come out, without putting it in secretly (...). All good theoretical physicists put this number up on their wall and worry about it.[4]

I'd like to propose that you look around in prestigious physics institutions, where many good theoretical physicists work, for these notes on the walls. I will give you 100 dollars for every exemplar you find hanging there. Feynman's spirit is long since dead.

DIRAC, THE PROTON, AND THE UNIVERSE

There is another mystery about the constants of Nature that has almost disappeared into oblivion. Apart from the fine structure constant, there is only one more possible way to combine fundamental constants so that all the units of measure cancel out and a pure number pops out. That's the ratio of electromagnetic and gravitational forces, lying in the range of 10^{40}. Yes, a number with 40 digits. But there is simply no reasonable math that produces a number of this order of magnitude. Every now and then, this vast difference in magnitude is discussed as the "hierarchy problem," but rest assured that in the realm of that terminology, you won't find any calculation that makes sense.

Without calculations, however, every attempt to solve the hierarchy problem remains just so much hot air. Dirac, who despised this and loved mathematics, wondered if Nature had come up with such a large number elsewhere. A few years after Hubble's discovery of the expansion of the universe, Dirac realized that its enormous size is about 10^{40} times the proton's size—the same huge number. Around 1938, after the first reliable density estimates of the universe,

Dirac found another surprising coincidence.[5] The number of protons in the universe is about 10^{80}, the square of that other tantalizing number.

Most notably, these values are islands on the number line, inaccessible by any other calculation you may be tempted to try with the constants of Nature. There is no fundamental property of the universe generating values such as, say, 10^{15} or 10^{50} or 10^{85}. Without being able to state a theory, Dirac said, "Such a coincidence we may presume is due to some deep connexion in Nature between cosmology and atomic theory." I am baffled by the fact that today Dirac's observation is mostly considered an irrelevant juggling with numbers. Physicists are skilled at playing virtuously with statistics when squeezing the last drop of signal out of a data mess, but they have lost any intuition for coincidences, about which you really have to wonder.

Pascual Jordan, who made important contributions to quantum mechanics, and who would probably have been awarded the Nobel Prize if he hadn't been a Nazi sympathizer, commented on the "large number hypothesis" in 1952: "I think that Dirac's thoughts are one of the greatest insights of our era, and their further examination is a main task."[6] However, in 1938, the dark signs of World War II were already casting clouds on science, and it was certainly not a good period for deep ideas to spread. Rather, the physics community had already felt the tension of the discovery of nuclear fission and foreseen its catastrophic consequences. Later, the gain in prestige and power from nuclear weapon development would lead to a completely new focus of fundamental research.

From time to time, Dirac, like an elder statesman, raised his reflective voice,[7] but particle physicists in the United States had long since taken the lead with their accelerators. Instead of thinking about unanswered questions, as Dirac did, they delivered many unasked-for answers in the form of a great heap of elementary particles, the remains of which we have discussed in the previous chapters. Though Dirac ventured a prediction on the variation of the gravitational constant, for which there is no evidence yet,[8] it is premature to dismiss this approach. Dirac's hypothesis is to date the only quantitative idea about the biggest problem of physics: how gravity relates to the microscopic world.

I speak of things known, for it is useless to speak of things unknown.

—*George Berkeley, Anglo-Irish philosopher*

MS. THEORETICAL PHYSICS: THE PROPHETESS OF HARVARD

Let us get back to modern theorizing. *Esquire Magazine* publishes a list of the "75 Most Influential People of the 21st Century." It is surely an exceptional coincidence if a person listed there has also authored one of the "100 Most Influential Books" selected by the *New York Times*. The name of the superstar of physics is Lisa Randall, a professor at Harvard, awarded with countless science prizes, at times the most cited physicist in the world. What did she find out that has brought her such acclaim?

Well, to make it short, the reason why gravity is as weak as it is, according to Randall, because of the small curvature of a fifth dimension. Any more questions? In Randall's model of the universe, this extra dimension is jammed in between two imaginary membranes ("branes") of a higher dimension.* We are sitting on one of the branes, called the weak brane, together with all of the matter and every force bound to our brane—just like water spots on a shower curtain, as Randall explains.[9] But couldn't it be that such water spots are the result of an overheated imagination?

Is this small curvature of the theoretical fifth dimension supposed to be an explanation of gravity? Why not in a sixth or seventh dimension? One could just as well explain gravity's weakness with pink elephants using their ears to shield it. As long as things are postulated as being invisible, such as the extra brane or the fifth dimension, every flash of genius is on an equal footing, but plainly untestable.

Undeniably, however, Randall's idea triggered a wave of research. In modern physics, how often one's paper has been cited by others has become a measure of an author's importance. But generating fashions does not mean that one has made it to the bottom of a puzzle. Compared to Dirac's thoughts on gravity, positing extra dimensions remains a lightweight approach, even if many more weak brains jump on the "weak brane" bandwagon. But, as the cosmologist Lawrence Krauss noted, "The as-of-yet hypothetical world of hidden extra dimensions had, for many who called themselves physicists, ultimately become more compelling than the world of our experience."[10] The mind goes belly-up.

EXTRA DIMENSIONS—PUTTING SKIN IN THE GAME

I believe in evidence.

—*Isaac Asimov, American author*

I almost forgot to mention it once more. All experiments searching for effects of the extra dimensions have come up empty handed. As they don't want to be accused of having lost grip with observations, theorists invent the fig leaf that one extra dimension may be big enough to be measurable.

---

*The name originates from the word "membrane." With all fantasies on strings already being discussed, the generalization to branes was inescapable.

If one still doesn't find anything, well, tough luck. But a closer inspection of the theory readily reveals that the extra dimension was just a bit too small after all, and the game continues.

The Eöt-Wash gravity research group at the University of Washington, mentioned in chapter 5, has been conducting gravity measurements at tiny distances, so that the extra dimensions should have revealed themselves,[11] but the dimensions seem to shrink as the experimental setup becomes finer. At the Marcel Grossmann Meeting in Rio de Janeiro in 2003, Ruth Durrer from the University of Geneva reported the results of a pulsar system, concluding that even one single extra dimension would increase the orbiting stars' gravitational wave emissions by 20 percent, which is in clear contradiction to the data. Immediately a question came from the audience: "Do you mean that all these theories are dead?" A tense silence spread through the room. Durrer constrained her death sentence to only a single class of string theories. "Why didn't you just answer with a simple 'yes'?" I asked, jestingly reproving her when we met two days later. In her Swiss accent, she replied, "Yes, perhaps. But they would have fussed even more."

But even when holding such a neutral point of view, any evidence for extra dimensions is just missing. Since absence of evidence never becomes evidence of absence for die-hard believers, the faith remains deeply rooted. Lee Smolin describes the incredulous amazement with which a particle physicist replied to his doubts about extra dimensions: "But do you mean you think it's possible that there are *not* extra dimensions?"[12] Martin Rees is said to have wagered his dog on the existence of higher dimensions, and the inflation theorist Andrej Linde has even bet his life. Gilbert Keith Chesterton once said that people are ready to die for any idea, provided that the idea is not quite clear to them. The extra dimensions are not quite clear to me, but I would rather act like Steven Weinberg. He would actually bet *Martin Rees's* dog as well as *Andrej Linde's* life on the existence of extra dimensions.

KNOCKING AT THE NUTHOUSE DOOR

The road to ignorance is paved with good editions.

—*George Bernard Shaw*

While discussing matters of life and death, we are touching on a subject that seems to make popular physics books even more popular—religion.

One professor of physics, for instance, has gone to extreme lengths to explain all the biblical miracles in a strictly scientific way in his book *The*

Physics of Christianity. The ability of Jesus to walk on water, for example, is attributed to a neutrino ray acting as a rocket booster. This reading is surely an amusement for the mentally healthy.

Lisa Randall's *Knocking on Heaven's Door* is far from such obvious nonsense; however, she feels it is worth mentioning the symmetries of the symbols of the world's religious communities.[13] That's not exactly surprising, but she then draws a parallel with the role of symmetries in modern physics, insinuating that both religion and physics shared some deep knowledge.

Ultimately, Randall has returned to Earth when publishing a book about the Higgs discovery, expressing little appreciation for the fact that diligent observers were yet unable to tell whether it was the Higgs or not.* As in the economy, however, goods may sell well before being produced.

Having an appreciation for solid evidence, I believe that science and religion are simply incompatible. At the legendary Solvay conference in 1927, where all the big minds of the time struggled over quanta, Paul Dirac argued, "If we are honest—and scientists have to be—we must admit that religion is a jumble of false assertions, with no basis in reality. The very idea of God is a product of the human imagination." Wolfgang Pauli, always ready to mock, teased him with, "Well, our friend Dirac, too, has a religion: God does not exist and Dirac is His prophet." Dirac laughed, but nowadays, some top physicists find symposiums like *Science Meets Religion* benedictory if not beneficial, when the claptrap is accompanied by some money on the table provided by the Templeton Foundation.

HOW TO WRITE BOOKS OUT OF NOTHING

Another book playing with semi- and pseudoreligious terminology is *The Grand Design* by Stephen Hawking, although it seems to be just a coauthor's rehash of the current fashion in physics, garnished with some babble that Hawking had never worked on. What all this does is expose how to squeeze as much as possible out of his name. Who seeks deep reflections on physics will find that the *Grand Design* is a grand delusion.

Hawking's disdain for religion is shared by Richard Dawkins in *The God Delusion*, and we must be thankful to Dawkins for debunking the Intelligent Design baloney with all its powerful relatives. And of course, while fighting the die-hards, the right-wing religious, and the hysterical sect

---

*Indeed, Wikipedia administrators were so annoyed at people claiming the Higgs discovery that there was an explicit warning when you tried to do this entry.

leaders, Dawkins needed at least a few allies. Understandably, he looked for them among physicists, but sometimes you can be wrong when choosing a run-of-the-mill cosmologist.

Dawkins rushed to write a neat afterword for Lawrence Krauss's book *A Universe from Nothing*, but wasn't he a bit afraid of talking too big when calling Krauss a Darwin of cosmology? Krauss is a funny character (and I really like his mockery about string theory), but he is just chewing the standard model's cud, as everybody else does. Krauss is a typical particle physicist who, after his field had been paralyzed by escalating costs, moved over into cosmology, equipped with wishful (and low-cost) thinking of the Universe as "the great accelerator in the sky."*

While Krauss tackles some unresolved questions of cosmology (besides those like Dirac's that he is not aware of[14]), his argument that the universe "comes from nothing" is plainly ridiculous. He does not spare the reader a single ludicrous story about the Big Bang era and ends up declaring eternal chaotic inflation as "the most likely possibility" for the beginning of our universe.[15] Let him like it. The real big irony here is that Krauss takes the starry-eyed attitude of the consensual canon—a true priest of modern religion. Dawkins, please write your next book!

FROM SHIVA'S DANCE ON CONDENSER PLATES TO QUANTUM FOAM'S NIRVANA

Unfortunately, mystic palaver frequently leaks into science. One popular parallel between "eastern" wisdom and "western" science was already constructed in the book *The Tao of Physics* by Fritjof Capra, one of Heisenberg's students. Being the student of a celebrity, however, doesn't guarantee that all your ideas are a flash of genius. Capra had an epiphany that the dance of the god Shiva is basically nothing more than quantum fluctuations, which, according to Heisenberg's uncertainty principle, allow particles to pop out of empty space by temporarily borrowing energy out of nothing (yes, this applies here).

Heisenberg's principle recently showed up in a beautiful experiment with two metal plates. In free space, virtual photons of every wavelength keep bouncing around, but in a narrow space between the plates, the longer wavelengths do not fit inside. As a consequence, slightly more photons hit the outside of the metal than the inside, resulting in a net force

---

*This phenomenon has been first described by the British astronomer Mike Disney (arxiv.org/abs/gr-qc/0009020).

attracting the plates. This is terrific evidence for quantum fluctuations.*
No Shiva needed so far.

Some theorists cannot help fantasizing that quantum fluctuations have
something to do with the Planck length of 10^{-35} meters, without any evidence of course. If that were the case, general relativity would then cause
extreme curvatures, "tearing the smooth space-time." Whatever that
means. Just like a stormy sea with its breaking waves, the fluctuations
would generate "quantum foam," a term coined by John Wheeler in 1955,
Einstein's last year of life. The calm, clear waters in which Einstein loved
to sail became a thing of the past. And so did the calm waters of scientific
reflection. Instead, we have been witness to the bursting dams of reason,
ripped and perforated structures, wormholes, and other nonmeasurable
sci-fi nonsense that has been pouring out of theorists' heads. No one knows
what it is supposed to mean, but *anything goes*. Physics has ultimately
turned into postmodernism.

WHERE QUANTUM MECHANICS CLASHES WITH REALITY

We can only wonder how many grown-ups deal with such theoretical toys,
since quantum phenomena that are actually testable are certainly interesting enough. Quantum mechanics can tell you the *probability* of finding a
particle at a given location, which is calculated by the so-called wave function. That idea won Max Born the Nobel Prize in 1954. In the strange world
of quantum physics, an electron may further be in a kind of schizophrenic
state called "superposition," which specifies the likelihood for its "spin" to
be directed up or down. In fact, quantum physics allows particles to live
several lives all at the same time. Yet the moment someone comes along
and tries to measure the particle, it jumps into one of the possible states.
This is a philosophically explosive statement because it implies that there is
no true reality independent of observation (and measuring is just another
way of observing).

Erwin Schrödinger, the "father" of the wave function, was disgusted by
this notion, and so he came up with a famous thought experiment. A cat in
a box is exposed to a poison capsule. When the poison's release is triggered
by a quantum process, Schrödinger argued, one has to assume that the cat
in the box is in a superposition state of both life and death, as long as the
observer doesn't open the box to see what's going on. Yet the argument is
an extrapolation from the behavior of one particle to the behavior of the

---

* Called the "Casimir effect."

cat's 10^{27} particles. Therefore, a debate about the poor animal is of limited benefit, prompting the annoyed Stephen Hawking to say, "When I hear of Schrödinger's cat, I reach for my gun."

MANY WORLDS AND EVEN MORE THEORIES

There is a somewhat more realistic way of looking at the cat-in-the-box problem. Even without humans observing, when a "choice" is made—in the cat's case, between life and death—the air molecules in the box would eventually put an end to the quantum jiggle between life and death. The cat, at least, would come to know its destiny. The American physicist Hugh Everett tried to explain this in an interesting way. Instead of believing that in every instant, Nature is randomly selecting one of the many allowed quantum states to be realized here and now, he proposed that all the other possibilities come true as well, but in alternate realities. This is the "many worlds" interpretation of quantum physics.

Based on Everett's idea, there is a universe in which Everett did not die early due to excessive smoking and drinking, and one in which his supervisor John Wheeler dissuaded him from publishing his "many worlds" thesis (this might be my favorite one), and so on.

Epistemologically, this solution does not do much harm, because quantum mechanics, while successfully describing the experiments, allows some fancy interpretation by theorists. Of course you may ask yourself what you actually learn about physics when theorists have been postulating about universes in which dinosaurs have survived and worry about the laws of physics,[16] and other such nonsense. I am happy with this universe in which I can stop listening to that stuff.

Some physicists seem to have added dope to Everett's sins while constructing their own theories. As with almost every idea in cosmology, parallel worlds may be perverted through combination with "cosmic inflation." This turns out to be especially useful for depositing absurd results that overstrain reality. "Eternal" inflation promotes the idea of new worlds being continuously created. Invisible ones, of course. It serves as an implicit justification to perpetually fantasize about grand new stories of the beginning of the cosmos.

The parallel worlds, called multiverses, are perfectly suited for the nasty question of why fundamental constants have the numerical values we observe. With a fine structure constant of, say, 5.3 instead of 137.036, carbon atoms could not form and neither could theoretical physicists. The explanation is that nature is producing invisible universes with all conceivable constants, and we happily live in one where life was possible. Of course, such arguments open the floodgates to arbitrary nonsense, as, for example, when theorists linked the fluctuations in empty space to the cosmological constant. Unfortunately the latter is too small by 120 decimal places! This could be called the wrongest prediction ever, but it does not harm the beauty of theory, when one cooks up an explanation using parallel universes.

PHYSICS, THE QUEEN OF THE IRONIC SCIENCE

Should these concepts turn out to be true, I shall not be ashamed to be the last one to believe.

—*Ernst Mach, German-Austrian philosopher*

There is a blooming of physical theories, but most of them have lost contact with experimental confirmation by many orders of magnitude. Instead of comprehending Nature, physics seems to see its purpose as generating new plots for *Star Trek* or *The Twilight Zone*. Theorizing becomes more and more a sort of displacement activity, where we don't want to realize our loss of reality. Robert B. Laughlin, awarded the Nobel Prize in 1996 for his work in solid-state physics, angrily criticized this attitude, launching the ironic proposal to describe black holes as "phase transitions."[17] When asked if this could ever be measurable, he replied, "Of course not, I'm just kidding. But I am fed up with sitting in seminars and listening to speculations about what happens when strings meet black holes. Nobody is talking about experiments."

For many of today's physicists, it has become a primary concern to see if more or less abstruse theories can be combined with each other. They debate whether a hypothetical particle called a "cosmon" is interacting with dark energy, whether black holes exist in two-dimensional gravity, and what kind of topology is required for wormholes in ten-dimensional space-time. But all these are splendid specimens of what John Horgan called "ironic science." In his book *The End of Science*, Horgan raised the right questions when interviewing practitioners of ironic science such as the inflation theorist Andrej Linde: "I asked if he ever worried that all his work might be—I struggled to find the right word—bullshit."[18] I disagree with Horgan's claim that science as a whole has come to an end, but there is no question that theoretical physics is sick.

Yes, our cosmos *may* have emerged pretzel-shaped out of a bubble multiverse. And the curved weak brane we may have next door may be just one of many extra dimensions. Among these hypotheses, there just might be the jewel leading to a grand unified theory. But we cannot sort out the gems from the garbage unless we have testable data. Physics is in an era of great narrators who see gold and jewels decorating Nature's garments. In light of these fabulous appearances, we should not be ashamed to take the children's role in this fairy tale. The emperor is naked.

There cannot be better service done to the truth, than to purge it of things spurious.

—*Isaac Newton*

Chapter 15

GOODBYE SCIENCE, HELLO RELIGION

STRING THEORY: HOW THE ELITE BECAME A SECT AND A MAFIA

String theory is the hypothesis that space-time consists of very tiny structures in the range of 10^{-35} meters. These one-dimensional strings (and their higher-dimensional analog, branes) are supposed to extend along additional dimensions of space-time. The vibrations of such strings are assumed to represent elementary particles. Initially, the theory aspired to calculate all the physical properties of particles, together with being able to explain the different interactions of physics in a unified picture.

String theory pretends to be nothing less than the ultimate unified theory of physics, the Theory of Everything (TOE), a unique accomplishment of human intellect. It is no wonder that the leading institutions of physics almost exclusively follow that road of research. Brian Greene's book *The Elegant Universe* explains why:

[A] unified theory has become the Holy Grail of modern physics. And a sizeable part of the physics and mathematics community is becoming increasingly convinced that string theory may provide the answer. From one principle—that everything at its most microscopic level consists of combinations of vibrating strands—string theory provides a single explanatory framework capable of encompassing all forces and all matter.[1]

And yet the theory is all-encompassing only to the extent that we define what the "everything" in a Theory of Everything is supposed to be. Shouldn't scientists at least know what they want to discover? The science author Gary Taubes expressed it like this: "The string theorists don't yet have a clue where this progress will lead them."[2] Is it really progress then? Undoubtedly, string theorists have revealed surprising mathematical theorems. Unfortunately, most physicists just don't seem to know what they're talking about.

Jeffrey Harvey, a professor of string theory at the University of Chicago, said, "The string theorists have however still to figure out what the hell it all has to do with reality."[3] "The answer might be 'nothing,' right?" wrote Dieter Zeh, a chair of quantum theory at the University of Heidelberg. Like many others, Zeh was irritated by the clash of the bold promises of string theory and the concrete results at hand—none at all, that is to say. Gerardus 't Hooft, the Physics Nobelist of 1999, hit the nail on the head: "Imagine that I give you a chair, while explaining that the legs are still missing, and that the seat, back and armrest will perhaps be delivered soon; whatever I did give you, can I still call it a chair?"[4]

PHENOMENAL PAINTING IN AN EMPTY FRAME

The physicist Lee Smolin tore the whole approach to pieces in his 2006 book *The Trouble with Physics*, arguing that string theory is basically lacking in everything—fundamental principles, consistent mathematics, and especially experimental testability, which is out of reach and therefore being replaced by vague hopes and speculations. Peter Woit, a mathematician at Columbia University, recently characterized the heavily promoted ideology of the last 30 years as follows: "We must have supersymmetry, so must have supergravity, so must have string theory,...so must have a multiverse where we can't predict anything about anything, thus finally achieving success."[5]

In his book *Not Even Wrong,* Woit judges string theory even more harshly than other critics. Woit's title refers to a remark made by the quantum pioneer Wolfgang Pauli, notorious for his sharp tongue. Pauli had categorized physical theories as "correct," "wrong" and "not even wrong," expressing utter contempt for the last type of theory, which was even worse than wrong due to its lack of testability. In the 1960s, Pauli derided an overly optimistic announcement of a theory of his friend Werner Heisenberg by sending a postcard to colleagues. It showed an empty frame with the caption, "This is to show the world I can paint like Titian. Only technical details are missing." Oblivious to the irony Pauli pointed out, many theoreticians nowadays speak of "frameworks" if they don't have any tangible results.

Mathematically speaking, string theory is, Woit said dryly, "the set of hopes that a theory might exist." The famous mathematical physicist Roger Penrose, though in a polite tone, also rubs salt into the wound: "How impressed am I that the very striking mathematical relationships...indicate some deep closeness to Nature?...I am certainly not convinced about it."[6]

Many practitioners of physics feel the same way, so that the Theory of Everything is sometimes called a Theory of Something or, still more to the point, the Theory of *Nothing*. And 't Hooft goes one step further. "Actually, I would not even be prepared to call string theory a 'theory' but rather a 'model' or not even that: just a hunch." However, most string theorists don't even have much of a hunch about the rest of physics.

THE SCIENCE OF THE REVOLUTIONARY GUARDS

Knowledge is concrete.

—*Bertolt Brecht, German writer*

At a legendary string conference in 1995, Edward Witten, indisputably the leading mind of string theory, suggested that certain versions of string theory might be "dual" to one another, and thus deep mathematical similarities may be inherent in them. This must have been an enlightenment for a great number of people, because everybody talked about the "second string revolution." The title of "first revolution" was reserved for the mere spreading of superstring ideas from the 1980s. I may be missing the punch line, but is the suggested assumption of a yet-to-be-defined similarity between two types of general hypotheses a result that could actually nourish physics?

And yet, more revolutions were to come. In 2001, David Gross, at the meeting of the American Association for the Advancement of Science, gave a talk entitled "The Power and Glory of String Theory," in a session called "The Coming Revolutions in Particle Physics." Three years later Gross would be awarded the Nobel Prize—though not for the announced revolutions. Maybe they just haven't been recognized yet. Or, the revolutions just lacked the results. I don't know.

Nevertheless these ideas have dominated the entire activity of the string community of the last decades. I believe 6,000 publications should speak for themselves (assuming they had to say something). Just savor that number like the quantum field theorist Bert Schroer did: "I don't know any non-metaphoric statement in the history of particle physics that would allow 6000 researchers to write an article about. You need a subject which is sufficiently vague and flexible to be massaged by so many people."[7]

As far as physics is concerned, these string "revolutions" are a storm in a teacup which has to date not poured a single drop of predictions. The lack of testability—once a death sentence for a scientific theory—is obvious. But the less the connection to reality, the bolder the aspirations. The Nobelist Sheldon Glashow noted, "Superstring theory is so ambitious that it can only be totally right, or totally wrong.* The only problem is that the mathematics is so new and difficult that we won't know which for decades to come."[8] Consequently, 30 years ago, string theory was already declared the yet-unexplored brilliant mathematics of the twenty-first century, a capital attribute that somehow has become less fashionable in the last decade. We are waiting for further superlatives to be developed.

> The gift to the 21st century which fell by luck already into the 20th century.
> —Edward Witten on string theory

MATH TOO DIFFICULT OR THEORISTS TOO FRIVOLOUS?

I sometimes feel that I should not criticize string theory "just" because it is not testable. The mathematics is really, really tough. To be precise, however, its toughness is also hypothetical. That's because to date, nobody knows anything about the equations to be solved one day, not to mention the theory's lack of fundamental principles, which makes it unclear what it can ever be based upon. Instead of getting the mathematics done, string theorists have instead turned to creating legends by carelessly citing incomplete results. Lee Smolin pointed this out in detail, mentioning, for example, the oft-repeated claim that string theory is "finite,"** an idea planted by an article in 1980 that relies on flawed proof.[9] Bert Schroer writes that string theorists created "a new culture of establishing a scientific truth starting from a conjecture and ending after several reformulations and turns with the acceptance within a community at the level of a theorem."[10] The entire reasoning has become, as Schroer pointed out in a series of articles, merely metaphorical.

Though not justified by anything, there is a widespread belief that elementary particles necessarily relate to string-shaped objects. Besides that vague idea, according to Sheldon Glashow, there are no convincing arguments in favor of string theory. He writes:

---

* One might, however, say that Glashow's standard model of particle physics always was happy with being a little right.

** This is considered an essential feature for a physical theory to make useful predictions.

Until string people can interpret perceived properties of the real world they simply are not doing physics. Should they be paid by universities and be permitted to pervert impressionable students? Will young PhDs, whose expertise is limited to superstring theory, be employable if, and when, the string snaps? Are string thoughts more appropriate to departments of mathematics, or even to schools of divinity, than to physics departments? How many angels can dance on the head of a pin? How many dimensions are there in a compacted manifold, 30 powers of ten smaller than a pinhead?

Glashow's concern about jobs has luckily proven to be unsubstantiated, because "also unfruitful ideas become immortal" if a sufficient number of people build their careers upon it, as Smolin scoffed. Even vague assumptions blown up into absurdity have their nutritive value.

DIMENSIONFUL THINKING—IT MAY BECOME SCIENCE

Despite the difficulty of string theory's mathematics, there are more than a few people who are not simply too stupid to understand it. Richard Feynman was one. He said: "For example, the theory requires ten dimensions. Well, maybe there's a way of wrapping up six of the dimensions...but why not seven? When they write their equation, the equation should decide how many of these things get wrapped up, not the desire to agree with experiment."[11]

Feynman summarized his argument rather tartly. "String theorists don't make predictions, they make excuses."[12] When he happened to meet the inventor of strings, John Schwarz, in the elevator at Caltech, he teased him with the question, "How many dimensions do you live in today, John?"

Putting the finger on that sore point of how arbitrarily string theorists play with dimensions is a pain in the neck for them. In his book *The Elegant Universe,* Brian Greene heartrendingly describes how the theorists Brandenberger and Vafa have recently struggled to justify why we live in a three-dimensional reality. It is five long pages, and ends in a trail of coulds, woulds, and might-bes, but allow me to refrain from summarizing. In popular books you may have seen the neat drawings of infinite surfaces, decorated with many small pellets, tiny cylinders, or even exotic pretzel-like-objects representing the extra dimensions. The less likely it is to come to know the exact fabric of these dimension carpets, the more intensely it is woven onto the theoretical looms since, according to Greene, it is almost dead sure that this is the "fabric of the cosmos."

In the world of string theory, the reader is guided through worlds with bizarre objects like o-het*erotic* strings, and Greene is not talking about a sex shop! For Greene soberly recounts Einstein's famous photoelectric effect (which doesn't work with red light...) and explains at length why electrons leave the metal with the same velocity.[13] You don't have to be familiar with it, but electrons don't do this by any means, as a high-school experiment can show. But maybe such niggling details don't apply to the extra dimensions where Greene seems to lodge.

METHODS OF STRING THEORY

Being a protagonist of string theory, Greene delivers some insights from his important contributions. A story of such revolutionary findings could, in theory, go like this:

> Talking to my dog, I had this idea that I initially dismissed as utterly lunatic. But at a seminar at Harvard I happened to tell it to Professor Deepthinker, who reflected on it for a moment, then raised his eyebrows and said: "This sounds highly interesting. You should take duality into account!" Having been encouraged by this, I immediately set to work with my colleague Jeff Loophater from MIT. Unfortunately our calculations predicted that every elementary particle must consist of 26 pink elephants, which was at first a very depressing result. But after great effort and developing a new renormalization method we managed to reduce this number to 11. One merely has to allow trunkless elephants, a result that was independently proven by John Fastscribbler and Jim Nitpicker from the Institute of Advanced Studies. Shortly afterward, Andy Wunderkind of Stanford University showed by means of a duality relation that the color could also be light blue instead of pink. These successes and the wide recognition we got very clearly indicated that we were on the right track....

String theorists, the real ones, tend to rejoice like little kids about the most marginal reduction of the absurdities they had invented just a moment before. They seem to argue that "we know all this seems utter nonsense, but believe us, it is an absolutely logical consequence of the nonsense we did before." Many accounts are written in such a fashion and make you wonder whether this is still Galileo's book of Nature, which, as the Italian heretic once stated, is written in the language of mathematics.

String theory has not, of course, been honored with a Nobel Prize,* and the science critic John Horgan has wagered $2,000 with the string supporter Michio Kaku that this will also not happen until 2020. Rather, string theory was acknowledged with a Fields Medal, the most respected award in mathematics. It qualifies a mathematician to be a physicist as much as a

---

* We will talk about the "Russian Nobel" in chapter 20.

Nobel Prize in Literature qualifies someone to be a politician. This would not be all too tragic, but at the moment, it seems that if physics were a democracy, we would say that elections have been suspended and the physics leaders are being appointed by a circle of literary critics.

THE LORD OF THE STRINGS

Forgive him, for he believes that the customs of his tribe are the laws of Nature.

—*George Bernard Shaw*

You can't understand string theory without Edward Witten. The early string theorists, Michael Green and John Schwarz, were considered stubborn, lone fighters in the 1970s and 1980s until Witten discovered them and brought them to the "Hollywood" of physics. It is said that Witten's phenomenal intelligence frustrated an entire generation of Princeton graduates working next to him, and his doctoral adviser, David Gross, ran out of math problems because Witten offhandedly solved all of them.

Witten's mathematical intelligence must be so all-encompassing that there seems to be not much left even for witty language. John Horgan gives an account of an interview with him. "This is another mark of the naïve ironic scientist: when he says 'truth' there is never any ironic inflection, no smile; the word is implicitly capitalized."[14] Dare Antoine de Saint-Exupéry say once more, that intelligence ruins the sense for the essential! What is more essential than the truth in physics?

Nobody will quibble about Witten's mental capacity. Even the string critics Smolin and Woit agree that they have never met a more intelligent person than Witten. Moreover, I have never heard anybody who had encountered Witten and *not* declared him the most intelligent person they had ever met. Brian Greene casually notes that Witten is regarded as Einstein's successor, as if we were talking about a new institute's chair. Yes, Witten is possibly the greatest genius of all time.

Of course, this also puts into perspective the casual encounters of Greene and Witten of which we are told. It is considered a great accolade among theoretical physicists if they are able to inspire Witten to contemplate a problem for one minute. I fear that 10 minutes of thinking might result in a Nobel Prize (once the Nobel committee eventually has recognized his capacities).

Witten's creativity is legendary, his productivity unmatched, and his intelligence unique. Those whom he listens to are anointed to a special realm, the books he lays his finger on are read by many, and his keyboard will no doubt someday be auctioned off on eBay. Witten the genius, Witten the new Einstein, Witten the greatest physicist of all time. The problem? He is not doing physics. Witten stands out among those who surround him. He may work faster, better, and be more creative than others, but as the Chinese proverb states, "The talent hits the target missed by everyone, but the genius hits the target no one has seen." There is nothing more to say about Witten and Einstein.

STRINGY LITURGY

Now let us venture into the unthinkable. Can an entire intellectual elite like string theorists have permanently devoted themselves to foolishness? Can string theory really be nonsense if the best and brightest are convinced of it? Yes, it can.

It's not the ability to rush through complicated calculations that we must consider, but rather a lesson from sociology and history: big brains can be brainwashed, as well as little ones. Take, for instance, the medieval theologians. Weren't they the elites back then? Didn't the brightest guys of that era stall science for centuries? They did so because the best minds of that time had no choice but to sharpen their intellect through the problems of medieval theology. Today's theoretical physicists are trapped in nearly the same way. A non-string theorist in a leading theoretical physics institution has become as exotic as an atheist in an Episcopal conference.

Unfortunately, the parallels don't end there. The wondrous stories in popular books on string theory mainly differ from creation myths in their time of origin. Just as the fascination with the godhood has its psychological origin, the fascination with strings is mainly a reflection of dreams and hopes. Physicists are served by string theories as humankind is by the world religions. It is a matter of faith.

What we are told is that there is a new era of mathematics our minds cannot yet grasp that will lead us to the Promised Land of strings. Those who delve into the subject in a like-minded community of experts—let's call them "monks"—are devoting themselves to abstruse calculation with great effort, an intellectual self-flagellation that is undoubtedly perceived as mind expanding. The absence of the gene of irony, as Horgan diagnosed for Witten,[15] also characterizes religion. Or, as the Swiss writer Max Frisch phrased it, "Given that you believe in God, is there anything that makes you believe he has a sense of humor?" American string theorists could not crack a smile when they were called a faith-based initiative. And indeed, it isn't a joke. String theory has long since become a religion.

TRACING TRACKS—WHERE WAS THE EXPERIMENT AGAIN?

Errors using inadequate data are much less than those using no data at all.
—Charles Babbage, English philosopher

Reviewing the evidence for string theory only confirms the impression of religiosity. The biblical gospels at least give a credible account of the feeding of the 5,000, while string theory just provides wondrous professors' salaries. "String phenomenology" is instead is a field for building careers, as being the head of the commission investigating the Eucharistic miracles. You may dispute how much string theory has entered that stage, but even grand unification sympathizers like the Harvard professor Howard Georgi are well aware of string theory's missing link to experiments and how dangerous that is:

> If we allow ourselves to be beguiled by the siren call of the "ultimate" unification at distances so small that our experimental friends cannot help us, then we are in trouble, because we will lose that crucial process of pruning of irrelevant ideas which distinguishes physics from so many other less interesting human activities.[16]

To wriggle out from such accusations, once in a while some fig leaves are provided that claim to be able to test string theory "in principle," "in the future" or with "indirect evidence," an abused term that helps to escape the here-and-now of observations.

Among all the predictions for experimental results from the Large Hadron Collider (LHC) results at CERN, the vaguest ones are from string theory, along the lines of "*if* the extra dimensions are large enough, one *might* see something." One of the most brazen after-the-fact predictions of string theory was hyped by Gordon Kane, the director of the Michigan Center for Theoretical Physics. Once all other energy regions were excluded as possible solutions by LHC experiments, and the first rumors started circulating late in 2011 that a signal at 125 gigaelectron volts might indicate the Higgs, Kane published an article[17] in *Nature* that concluded that it was precisely this value that supports strings!* At this point, you may ask what it actually means to hold a theoretical physics chair in a prestigious university if you can just follow your fancies into the realm of fantasy.

---

* I forgot to add he was awarded a prize by the American Physical Society for it, www.math.columbia.edu/~woit/wordpress/?p=4553.

Another sport practiced in string theory is predicting *violations.* In this game, you take any good theory—special relativity, laws of energy conservation, or whatever—and any deviation from it, no matter what, where, and how, is deemed to be a triumph for string theory. Again we note that religion starts where science ends.

> Theoretical ideas discussed must be supported by experimental facts. Neither supersymmetry nor string theory satisfy this criterion. They are figments of the theoretical mind.
>
> —*Martinus Veltman, Nobel laureate*

OTHER SUCCESSFUL POSTDICTIONS

> Predictions are difficult—especially for the future.
>
> —*Niels Bohr, Nobel laureate 1921*

Do we really need particle colliders inaccessible by today's technology to test string theory? Actually, there are some particles that we *can* study without resorting to such high energies, such as the ones we consist of. It is therefore a misstatement that string theory does not have any experiments at hand with which to establish its glory. More than 20 values in particle physics, mostly masses, are awaiting an explanation. Frank Wilczek, the Nobel Prize winner in 2004, said: "The early promise of superstring theory to calculate these quantities has faded after decades of disappointing experience in attempts to construct phenomenologically adequate solutions, together with the discovery of multitudes of theoretically unobjectionable but empirically incorrect solutions."[18]

The failure to explain masses is now bizarrely being oversold by the claim that string theory "reproduces the standard model." That's like claiming to draw a beautiful straight line and starting with a lot of bumps.

But let's focus on the big picture. Instead of grappling with these tedious masses, Edward Witten put forward a revolutionary insight— that string theory predicted gravity! The moment at which this occurred to him, he said, was the intellectually most satisfactory one in his life. Later on, he even predicted the existence of the proton. The string community appeared spellbound by these epiphanies, but gravity is not at all a mystical experience. We can measure it with a gravimeter. Notice to string theorists: please come back down to Earth! It would be slightly more satisfying if string theory could predict something we don't already know about.

SAY A LITTLE PRAYER

For ideas that hardly seem to have any real meaning, such as "string cosmology," Wikipedia offers this: "String cosmology is a relatively new field that tries to apply equations of string theory to solve the questions of early cosmology." This actually describes one of the bogus science subsidiaries of string theory. It opened after the speculation bubbles of its main enterprise of particle physics deflated.

I once asked one of these conference-to-conference salespeople for string theory, who had given a talk entitled "String Theory and Cosmology," about predictions. The question was if we should now expect—in the same vein as Witten's discoveries—that string theory will soon have predicted the Big Bang. "I can imagine what you want to say," he replied, "but despite Smolin, I still believe in the predictive power of string theory." Yes, he believes! Sustaining one's faith is such a difficult thing today. In order to emerge strengthened from such blasphemous attacks as the books of Woit and Smolin, I recommend you cross yourself in 11 dimensions and repeat the following cleansing gospel three times.

21st Century Lord's Prayer
Our Witten who art in Princeton,
published be thy name.
Thy brane come,
Thy theory be done,
On earth as in other dimensions.
Give us this day our daily idea,
and forward us our progresses,
as we do forsake the path of science,
and lead us not to contemplation,
but deliver us from observation.
For thine is the string,
and the power, and the irrefutability,
for ever and ever
Amen.

THE VALUE OF SECULARIZATION

No more mercy with those who haven't explored but talk nevertheless.

—*Bertolt Brecht*

The crudest method of escaping the principles of empirical science is, of course, to simply discard them. Since string theory doesn't provide any testable predictions, we have to get rid of the apparently old-fashioned, strict

limitations of scientific methodology, right? But there are much more subtle ways to get around the rules. Instead of launching a barefaced attack on
Karl Popper's idea that scientific theories must be falsifiable, some scientists
are tricky enough to quote Thomas Kuhn, another philosopher of science
and a critic of Popper. According to Kuhn, science undergoes paradigm
shifts like the transition from the geocentric to the heliocentric worldview.
Here we are: Strings are an epochal paradigm shift that must leave behind
the testability of "old physics"![19] Pope Edward Witten is transformed into
a Galileo-like heretic by crooks wearing police uniforms. Witten himself,
in a papal manner, is "reflecting on the fate of spacetime."[20] That's certainly
safer than reflecting on the fate of string theory.

String theory is beautiful. Extraordinarily beautiful. At least that's
what everybody says, so I won't bother you with my supposedly underde-
veloped sense of aesthetics. It is, however, extraordinary for science that
beauty has become a self-contained argument. String theory is just too
marvelous, too beautiful, to *not* occur in Nature. In the eleventh century,
Anselm of Canterbury had "proven" the existence of God in much the
same way. Since God possessed all conceivable qualities of greatness, a
nonexistence would be a lack of at least one of these features and result
in the being not being God. Anselm is regarded as the founder of scholasticism. He should actually no longer be ignored in the history of string
theory.

IRRATIONAL EXUBERANCE: DYING AFTER THOUSANDS OF SUPERLATIVES

It is remarkable how coordinated is the herd of string theorists who attack
the same problems, led, of course, by their prophetic bellwether Witten.
Roger Penrose was not the only physicist to observe that "wherever Witten
goes, the rest will follow soon." Julian Barbour remarked subtly, "Science
has its fashions. String theorists are a bit like a pack of hounds following an
extremely promising scent. But it is a particular scent. If they lose the trail,
nothing will come of the great chase."[21]

If string theorists do lose the trail, it will be the greatest waste of intelligence in history. There are only two possibilities. Either string theory
will be the ultimate unified theory, or it will be the biggest load dropped
into the garbage dump of science since Ptolemy's geocentric universe that
dominated science for 1,500 years. Let's hope that it won't take that long.
"One day we may understand what string theory truly is," said Witten.[22]
Sometimes you just need to understand a little history.

HOW TO BECOME HISTORY WITHOUT A STORY

Thy is to maunder, since you are the bell of your actions.

—*Friedrich Schiller, German writer ("Mary Stuart")*

Historians of science have often criticized Einstein for devoting 20 years of his life to a theory (unification of gravity with electromagnetism) that is considered a dead end. He would have been better off going sailing, was a biographer's posthumous counsel.[23] What is considered bullheadedness in one person turns into something celebrated if thousands of researchers follow that bullheadedness for decades. String theorists are now becoming historians of their own failed ideas. The first "I was there" papers are beginning to appear. Take, for example, the former director of the Aspen Center for Physics, who bravely outed himself by publishing an article entitled "Memoirs of an Early String Theorist"[24] about the "intellectually most satisfying years of his life." It's quite an accomplishment to stay satisfied without results.

One of the nicest circular arguments floating around is to say that string theory is "the only game in town." Oh, we say sadly, there are no reasonable alternatives to string theory! Nathan Seiberg, Witten's collaborator, announced, "If there is something that goes beyond, we'll call it string theory." This wasn't a joke! I would call it a form of pantheistic assurance, where "all" is comfortably and assuredly string theory. This kind of intellectual arrogance has, unfortunately, been corroding the credibility that true science relies on to survive. From a historical perspective, one has to reflect on how physics can get rid of string theory. Maybe one needs to establish a faculty for metaphorical mathematics or creative writing with mathematical methods. The problem is, as recently Jim Baggott put it, that string theory is still labeled as science.

The public should be more patient about the truth.

—*Lisa Randall*

EUPHORIA BECOMES CENSORSHIP, NOT AN INNOCENT GAME ANY MORE

Freedom—the first-born daughter of science.

—*Thomas Jefferson*

Paul Ginsparg, a string theorist from Cornell University, has the merit of having developed the platform ArXiv.org, which makes scientists' research openly available to anyone who wants to see it. But he has unfortunately also transformed it into one of the worst monopolies of the Internet. There is no way for a serious researcher to get around ArXiv. Try not to get on their blacklists. Although it is denied officially, some researchers are blocked automatically from publishing there. Ginsparg has fought ferociously[25] with those who put forward papers that are not part of mainstream science, while a mountain of string papers was piling up on ArXiv. He appears to be a cop ticketing nose pickers while sitting on a big heap of shit. ArXiv has liberated itself from its founder, but the spirit of the moderators persists. Many of them are string theorists shifting disagreeable papers to low-rank general categories or even banning their authors. Cornell is a shame for a country where the freedom of speech has a value.

Needless to say, the dominance of string theory makes it hard to get alternative ideas published in mainstream journals. But everywhere in theoretical physics you have to deal with string theorists. They are still in power, having withstood the storm brought up by Smolin and Woit by waiting it out like politicians in a scandal. Scientists have to get their proposals reviewed, and you have to be careful to formulate the theoretical aspects. Therefore, many researchers kowtow to the string merchants with clenched teeth, in order not to jeopardize their funding. Casually mentioning that some phenomenon under investigation "might be one day solved by string theory" has become the tip of the iceberg in a sick, dishonest game.

FROM THE SPIRITS THAT I CALLED, SIR, DELIVER ME!

Men, it has been well said, think in herds; it will be seen that they go mad in herds, while they only recover their senses slowly, and one by one.

—*Charles Mackay, Scottish poet*

There is no suggestion here that Witten or others went into string theory with any intention of causing harm to science. But after a generation of theoretical physicists employing their talents to advancing string theory, we have arrived at a point where there are 10^{500} different versions of string theory, and there is no rule whatsoever that guides us toward a reasonable choice in this chaos. This arbitrary construction is called "landscape," a perspective Witten found utterly discomforting, so much so that some feared his retreat from string theory altogether. Imagine the pope leaving the church! However, there is a so-called anthropic "solution" to the problem of so many possible string worlds. One of this nearly infinite number of universes (multiverses) must be realized—because otherwise we humans wouldn't live here in it!

David Gross fights against this multiversion absurdity with Winston Churchill's words: "Never, never, never give up!" Gross has become aware at this point, that the theory has slid into pseudoscience, though he and his "Princeton String Quartet" have contributed a lot to this sad end. Just like a confused mountain guide, Gross is lost in the new "landscape" of prophets such as Leonard Susskind, who don't look into the gorge but continue to promise the pie in the sky. It seems as if string theory is providing a sort of negative evolution, where the most stupid ideas prevail.

What next? In YouTube videos[26] you can follow the birth of future string theories. Witten sketches a line on a piece of paper, connecting the five conventional string theories, and that's about it. Oh yes, a name is invented, "M-theory." As Witten explains, it refers to *magic, mystery,* or maybe *matrix.* Brian Green would presumably call it the *mother* of all theories, but there are also mockers such as Sheldon Glashow, who see the "M" as an upside down "W" for Witten. The cheeky João Magueijo associated "M" with mathematical masturbation. I do not want to pursue this idea, but be careful. Theoretical physicists have become frustrated with their eternal fiancée, the standard model. String theory is the hot mistress, whose thrilling sexiness outstrips all imagination. Up to now, however, it's all in their minds.

Part IV

BACKING UP

Chapter 16

CLEAR WATER

REASON VS. CIRCULAR LOGIC: HOW SCIENCE SHOULD WORK

One of the major topics discussed in popular books on string theory is the anthropic principle, a mysterious selection rule that is supposed to be responsible for the values of the constants of Nature. Fred Hoyle, an astrophysicist who had coined the concept behind it, would surely not be pleased with today's (ab)use of the term. Hoyle had wondered about the origin of chemical elements in the early universe and convinced the nuclear physicist William Fowler to perform an experiment with carbon, for which the Nobel Prize was later awarded... to Fowler. To back his conjecture, Hoyle had humorously argued that he had observed physicists (who in part are made of carbon) in our universe.

Of course, the constants of Nature must have made it possible for life to originate, because it obviously has. This self-evident fact is called the "weak" anthropic principle, which was then generalized into the nonsensical "strong" anthropic principle. The strong principle says that the fundamental constants have the values they have, *because* life wanted to originate through these. This perverts the interplay of theory and observation so that predictions—the essence of science—turn into postcognitions: things are as they are because they were as they were. The Nobel laureate Burton Richter said about the anthropic principle: "I think, it is one of the most stupid ideas ever to infect the scientific community."[1]

WHY DO ALL OBJECTS FALL DOWN?
THOSE FALLING UPWARD ARE ALREADY GONE.

Some values of the constants of Nature appear to be essential for the existence of the universe (e.g., for the existence of stars), though they seem to be as unlikely as rolling the dice and getting a "six" ten times in a row. In that case, it's not unreasonable to suspect that the dice is biased, meaning that we haven't understood the laws of this particular dice yet. The proponents of the anthropic principle would, however, say that the dice must have already been thrown a great many times in order to yield such an improbable result. Applied logic has a rather harsh name for this conclusion: inverse gambler's fallacy.

String theorists apply this to their "landscape" of 10^{500} versions of the theory, since no one can explain why one specific universe is singled out. Keen thinkers such as Leonard Susskind inferred that the choice was made in order to make life possible. *Homo sapiens* is quite full of himself here, but this regression into magical thinking is spreading in the string community. Recently, a landscape enthusiast noted[2] that the anthropic principle is restoring the special role of humans in the universe, which is not far from saying that God has chosen to make humans special. Maybe the anthropic versions of string theory will soon merge with Intelligent Design as advocated by Abrahamic creationists to crusade against the last bit of science in the Western educational deserts.*

In the case of the strong anthropic principle, the lack of content is not hard to see. Some call it the "completely ridiculous anthropic principle," or CRAP.

FAIR PLAY VS. EMPTY CHIT-CHAT

Science's greatest gift to civilization is its acknowledgment of fallibility.

—*John Polanyi, Nobel laureate*

It is worthwhile at this point to reflect upon the characteristics of science more thoroughly. Among nonscientists, there is a widespread preconception that science is concerned with what is "provable," but as a matter of fact, the opposite is true. At the core of a sound scientific statement is the acknowledgment that it could also be wrong. This is fair play. You have to specify an experimental result that would demonstrate that your results *might* be wrong.

---

* My emotions are somewhat at play here, since my students occasionally land up in such places, when doing an exchange year in search of the big wide world.

The philosopher Karl Popper considered the criterion of falsification—that means you *can* be proven wrong through experiment—essential for a theory to be scientific. A textbook example for such nonscience would be making theories about the interior of a black hole, because it is defined as a place where no information can possibly escape. And how would Popper feel about dark matter? Some have claimed it might have gravitational effects *only*, assuming that it doesn't emit *any* radiation. Savor this. How can you possibly be proven wrong, if you are predicting that nothing can be seen?

In these cases, the lack of evidence becomes the evidence, and we are dangerously close to the border between science and belief. Such concepts cannot be demonstrated to be wrong, so in practice, they are unfalsifiable. It is therein that the scientific death sentence lies.

When it comes to any theory where progress is stalled because we are lacking experimental tests—and hence any possibility of falsifiability—I have more interesting things to do! Should I have expressed myself in a mistakable manner in the previous chapter on string theory, I shall clarify. String theory is not hokum, no! There are merely more interesting things to study.

PHILOSOPHY OF SCIENCE, OLD AND NEW

We do not want to flog to death the discussion of Popper's criterion of how to define science. Other philosophers such as Thomas Kuhn and Paul Feyerabend have made valuable contributions to defining good science as well. Humorlessly haggling over philosophical questions soon becomes deadly boring. Richard Feynman may well have been referring to this when he said, "Philosophy of science is about as useful to scientists as ornithology is to birds." However, an ornithologist of any school of thought, whether it be Popper's, Kuhn's, or Feyerabend's, can help in cleaning the feathers when an oil spill like, say, string theory occurs. It would be great if the theoretical masterminds of our time would inform themselves about the ideas of any of these thinkers.

Today's philosophers of science are seldom taken seriously by physicists, probably due to their lack of solid knowledge in physics. On the other hand, the philosophy produced en passant by today's physicists sometimes appears dilettantish as well. In all seriousness, an MIT cosmologist claimed that the universe is not only described by mathematics (a famous, though not too humble, statement by Galileo) but that it *is* mathematics, with its symbols unchained from the baggage of meaning... and thus, utterly meaningless. This whole philosophical concoction is entitled "Shut up and calculate."

I don't think a science that tells its members to hush will advance very much. Moreover, the problem of today's physics is that there is way too much calculating and too little thinking. I am not willing to develop a new philosophy of physics, but instead of "Shut up and calculate," I would recommend: "First think, then speak."

THE LOGIC CIRCULATING IN THE GREAT MINDS

Nonfalsifiable theories, according to Popper's definition, can begin the chain of logical deductions just about anywhere and never come into contact with an experiment. There are also neat examples, where arguments form complete orbits, so-called circular reasoning. A relatively harmless case of circular reasoning is Lisa Randall's argument about undetectable higher dimensions. "Additional spatial dimensions may seem like a wild and crazy idea at first, but there are powerful reasons to believe that there really are extra dimensions of space. One reason resides in string theory...."[3] Oh, I see! Extra dimensions should exist, because string theory *requires* higher dimensions to exist. This is a classic circular argument.

There is another widespread "logic" that requires debunking—that the "beauty" of a theory is an indication of its correctness. Human beings appear to have a hardwired preference for symmetrical patterns, which appear beautiful to us. And physicists look for symmetries because they assume that the laws of Nature rest upon these. To sum up: the theory is correct because it is beautiful, beautiful because it is symmetrical, and symmetrical because symmetry was our prerequisite for being correct.

Usually, circular reasoning rests upon such hidden assumptions. As Immanuel Kant pointed out when he dissected Anselm of Canterbury's proof of the existence of God, while defining "perfection," the existence of God was already presupposed.

A BEGINNER'S COURSE IN SCIENTIFIC LOGIC

There is a paradox that theoretical astrophysicists use to fool themselves, which traces back to the philosopher Carl Gustav Hempel. Let's consider the statement "all ravens are black."[4] Now, to be strictly logical, consider an equivalent claim that "everything that is not black is not a raven." However, things become a little more complicated as you look for supporting evidence for this "law of Nature." Observing a black raven surely speaks in favor of the validity of this law but, by the logic of the equivalent statement, a white shoe would also be evidence that all ravens are black because it is not black and not a raven. Our common sense will suspect that something is wrong here, because a white shoe would also provide evidence for a "law" that "all ravens are yellow." As scientists, we might agree that we if we want to learn something about ravens, it is advisable to observe ravens.

By the same token, the nonobservation of something presents a logical pitfall because it allows us to conclude literally anything. That's what is happening when interpreting the nonobservation of magnetic monopoles, seen by Alan Guth as support for inflation theory. Sorry, but the nonobservation of such cloud castles just favors one conclusion—that they don't exist. I am still stunned at how Guth could impress the entire elite of US universities with this raven-black logic. Nonetheless, the fantasies supported by nonobservations are continually spreading.

OCCAM'S RAZOR HASN'T BEEN SHARPENED FOR QUITE SOME TIME

Observations are indispensable for science, but solid science has another characteristic—a desire for simplicity. The fewer parameters (numbers fitted to describe the data) a theory needs, the better. This becomes obvious if you compare Ptolemaic astronomy to Copernican astronomy.

The description of the data provided by epicycles wasn't that bad, but dozens of free parameters were necessary to accomplish this, instead of just one parameter in Newton's theory, the gravitational constant G. This is where the famous "Occam's razor" comes into play, a principle named after an English philosopher of the fourteenth century, which states that when two theories compete, the simpler one should be preferred: *Pluralitas non est ponenda sine necessitate*. Ockham's criterion also tells us that we should only trust sufficiently simple theories to prevail in the long run. The fact that there are a huge number of parameters in the standard model of particle physics is alarming, and cosmology is following in its footsteps toward increasing complication.

We sometimes hear that it is possible that Nature isn't that simple. Others argue that it is hubris for humans to want to understand everything. No, sorry, it is the very business of the theoretical physicist to find simple explanations of a plethora of phenomena. And thinking that Nature is horribly complicated, rather than that humans are too dumb to understand it as of now, isn't all that humble either. Occam's razor is as essential to science as the scalpel is to surgery. It may hurt. But if you don't want to pose the question of simplicity, you're better off reading the horoscopes.

POPPER LITE: LEGALIZING THE IDEOLOGY

A book emphasizing nonfalsifiability, such as Peter Woit's *Not Even Wrong*, is a pain for string theorists, and it shouldn't be a surprise that Karl Popper is not particularly liked in the string community. Leonard Susskind disdainfully nicknames Popper supporters "Popperazzi." Emperors of physics, such as Leonard Susskind, understandably have an aversion to Popper because he exposes the emperor as naked, threatening the sales value of his wonderful clothes.

In the end, however, ideologies such as string theory are the easiest to identify, since they have so completely lost touch with reality. There are others who give a nod to Popper's falsifiability, but then suggest it should maybe be loosened, relativized, or modified.

Besides the issue of absolute nontestability of various theories, there is also a practical nontestability at the Planck length of 10^{-35} meters. Such entirely hypothetical experiments conducted in the far future, if ever, make me rather suspect that theorists are actually quite comfortable being left alone in their experimentally inaccessible playground. Should we loosen Popper's dictum for these guys? No! If you are not planning on actually discovering something in your life as a researcher, then you should get out of the way of those who do. Firstly, there are plenty of organizations where people predict the future and preach about the afterlife. Secondly, it is utterly naïve to believe that anybody would care about our present theories in a distant future of divine measuring accuracy. And whoever is truly convinced of his or her brilliant but untestable idea just has to deal with departing this life as an unrecognized genius. People take the present, and themselves, far too seriously. Whether an experiment that might falsify an idea is theoretically or just practically impossible, the consequence must be the same—so long, theory! All we can do is wait for the evolution of new measurement techniques, and maybe explain to our fellow inhabitants how to preserve the Earth for the next decades. Otherwise we could bid our goodbyes to these theories and their tests anyway.

INJECTION OF FRESH EXPERIMENTS INTO THE THEORY BALANCE

Scientists are laughing their predecessors, but few realize that someone will laugh at their believe in the (disappointingly near) future.

—*Nassim Taleb*[5]

Even more treacherous to good science are those theories that every now and then lay out a lure that can, in principle, be reached by the currently available technology. However, it often turns out that the existing equipment does not quite suffice, but the one being financed for this alluring theory probably will, maybe... Right now, new far-fetched fantasies are developing that should justify colliders with still higher energies. The next generations of colliders will be capable of... feeding the next round of wishful thinking.

It is, by the way, also naïve to think that the standard model of elementary particle physics can still be falsified, in the Popperian sense. I expect that some initially measured "nonstandard" features of the Higgs boson will be transformed into evidence supporting a "standard" Higgs during the next couple of years. It will be based on the argument that there are "no reasonable alternatives" for the Higgs, so this must be it.

Just like the big banks in the economic crash in 2008, the standard model has long become too big to fail, and it is gobbling up taxpayer money in a figurative and literal sense. Firing up a new particle accelerator is the equivalent of an infusion of fresh capital that triggers a speculation market boom. The problem is that the dumb brute-force experiments of the standard model create the illusion of scientific progress. Whatever signal is detected, no matter how much it disagrees with the expectations (assuming there are any), is praised as a big discovery. No doubt the addition of every epicycle to geocentric astronomy was also praised as a big discovery.

Dinosaurs like the standard model erode science slowly but thoroughly with their effective nonfalsifiability. This creates a triple threat for science: vast resources are being appropriated, draining other fields of physics; a sick theoretical model is being kept alive; and a heap of new, poorly understood data is piling up.

SIMPLIFY YOUR SCIENCE: WHAT, WHERE, HOW, BUT FIRST—HOW MUCH?

God created everything by number, weight, and measure.

—*Isaac Newton*

A scientific theory can show progress if it makes numerical predictions and is thus quantifiable. In any case, the need to respond to the question "How much?" would make those who are used to escape to higher energies bite the bullet. So dear theorist, please tell me: What is the precise mass of the particle you are predicting? Will you admit defeat if it's not found within the error margins? I fear that this radical cure would make most models crumble, and thus the theorists, already driven by self-preservation, will invent an infinite array of excuses for not putting their predicted values on the table.

Again, I'd like to be clear that I am not against new experiments. But we should label them for what they currently are—giant, expensive toys we want to play with. It is fair to ask what fundamental insight, aside from all the gimmickry, has been delivered to science.

Being quantitative also means not trying to beautify a theory with qualitative ad hoc assumptions. Why, for instance, is dark energy now suddenly being said to be repulsive? Or better yet, why can't isolated quarks be observed? The quark "inventor," Murray Gell-Mann, asked the experimenters to look for single exemplars of quarks, but simultaneously upheld a neat loophole. The nonobservation of single quarks, he assured, would prove the supposed nonseparability of quarks. He was mocked by a colleague: "If quarks are not found, remember that I did predict that, but if you detect them, keep in mind that it was me who proposed the search."[6]

YE SHALL KNOW THEM BY THEIR WORDS

For mathematical equations, there are only two possibilities: either they are correct, or wrong. For models, there is a third one, namely, correct but irrelevant.

—*Wolfgang Pauli, Nobel laureate 1945*

At first glance, it is not easy to distinguish publications with a reasonable link to observation from those that merely make vague references to them (some have good reason to be vague). Luckily, scientific language has some identifying marks that allow us to track where soft theories skirt around the hard facts.

Follow along as we take a quick romp through the dictionary of scientific bafflegab. Calling a theory a *model* means that it is considered temporary and doesn't have the ambition to explain much of anything. It has no place in fundamental science in the long run. The same holds for a *class* of theories, which is never proven wrong by the failure of its members. Thus, many modern theories supply *tools* and *skills*, which usually means that we have tools but we're not using them. If an approach is said to *lead the way* or *prepare the ground*, then you can be sure that it will continue to do that for some time, even longer if *hopes rest* upon it. What is at rest here is progress.

If you read that a view is *substantiated*, then it must have been just a dream prior to that, whilst *promising* models are not actually expected to deliver on those promises, and *candidate* theories don't even know what to become. When a theory *carries the seeds*, most likely weeds will grow out of it, and if it is *on its way* or *well established*, this just means that the sickness is widespread; if it is *rich in opportunities*, it is just poorly supported. Then one can always *advance understanding*—this is still possible even when an idea is too advanced for a sane mind. Finally, if a research group *sets out* or *heads off*, it means that they don't have a conclusion, probably not even an objective, apart from securing their funding from the National Science Foundation.

Wherever *signs are pointing*, they mainly go nowhere, and the words *toward*, *trends*, and *recent developments* are misleading terms conveying that nothing specific, much less quantitative, is being predicted. At best, they describe so-called *frameworks*. Even if a tangible experimental result is utterly absent, a researcher can still read *footprints* from the data, while *signatures* indicate that the body of evidence is missing and there aren't even any footprints. And a really hazardous situation for the mind is if data is *compatible* or *consistent* with a theory. Apparently it is unknown to many that no false statement can be deduced from a correct one, but surely vice versa—this is elementary logic. Sadly, it implies that even the most arbitrary nonsense can predict, after many theoretical rinse cycles, that water is wet. A textbook example of this is cosmic inflation, where some laundry operators see inflation as proven by the cosmic microwave background. It is *compatible* with inflation,* but only after a huge serving of incompatible logic.

I KNOW THAT I KNOW NOTHING. AND NOT EVEN THAT (KARL POPPER)

At every stage we will think we understand, but at every stage there will be nagging doubts in the minds of those who wonder.[7]

—*Margaret Geller, American astrophysicist*

Aside from all the talking about it, will people ever discover a unifying theory, as has so often been stated? Roger Penrose, after his 1,000-page journey through physics in his book *The Road to Reality*** does not believe we are that close to its discovery. And who of the current "masterminds"

---

* Not to mention the recent LHC results, which are reported to be *consistent* with a Higgs boson, as if that would mean anything.

** People like to discuss whether we have come *closer* to the truth. Feynman snubbed such an interviewer: "If knowledge is limited, and today we know more than previously, of course we are *closer*. But what a stupid question!" (quoted in Gleick, 353).

should identify the theory and how? With an eye to the bold promises of theoreticians, David Lindley wrote with trenchant irony:

"There is no way to know in advance what this theory of everything will look like, but most physicists like to think when they see it they will recognize it."[8] There are serious indications, however, that the beaten paths are not leading to the great goal. We can perhaps learn more from the ancients than from modern prophets; the heliocentric worldview, for example, was anticipated by Aristarchus of Samos around 250 B.C.

It is sobering instead to see how many times in history we've been convinced of having reached the end of physical knowledge. In 1894, Albert Michelson (who later won the Nobel Prize) said: "The more important fundamental laws and facts of physical science have all been discovered, and these are so firmly established that the possibility of their ever being supplanted in consequence of new discoveries is exceedingly remote... our future discoveries must be looked for in the sixth place of decimals."

Today we would laugh at such a statement, but in modern physics we have seen a similar exuberance. Stephen Hawking announced around 1980 that the theory of everything was at hand and the end of theoretical physics was "in sight." This end could be a dead end, too. History wouldn't care very much. She has seen this happening a couple of times.

Once again, Karl Popper seems to be a bit wiser: "We know nothing—that is the first point. Therefore we should be very modest—that is the second. That we should not claim to know when we do not know—that is the third. This is more or less the approach I should like to popularize. It does not have good prospects."[9]

Popper was expressing a fundamental characteristic of scientific thought—skepticism. In contrast to euphoria, this is a very individual feeling. Accordingly, in the herd mentality of today's Big Science, there are no big doubts about whether we are on the right track with our models, and the underlying concepts are questioned even less. It was, however, the skepticism of Copernicus, Galileo, Faraday, and Einstein that advanced science by challenging the conventional wisdom, not the celebration of the alleged successes of the standard models, whether at the time of Michelson or today.

A return to the successful principles of physics is urgently needed. Testability, logic, and simplicity are the "liberty, equality, and fraternity" of good science. And just as politics ought to have certain morals, science cannot do without skepticism. But we are often taken for fools.

Theoretical Physics:
The Take-off to Fantasy

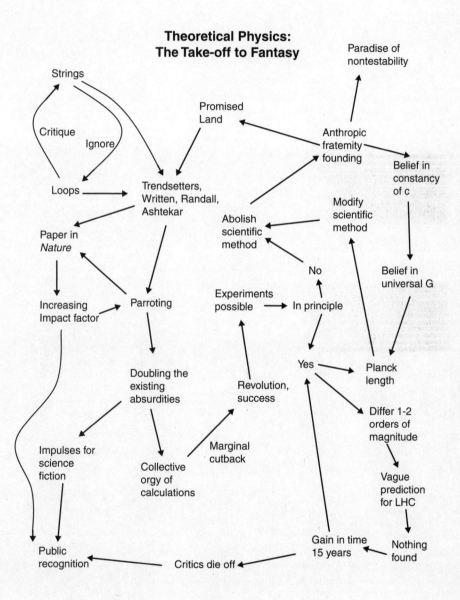

Chapter 17

WELCOME TO BYZANTIUM

COMPLICATIONS ON COMPLICATIONS: HOW PHYSICS BECAME A JUNK DRAWER

Karl Popper's critical rationalism is a sharp knife we can use to distinguish between real science and its fictional knockoffs. A bit of healthy skepticism, the requirement of falsifiability, and the need to quantify results are all invaluable instruments for diagnosing theories. Scientists should never allow them to be watered down by dreamers, boasters, and Big Science lobbyists defending their prestige experiments. This toolbox of critical rationalism provides the rules for how science ought to work, and it holds up a mirror that reflects back all the obvious sicknesses of science.

Popper's idea of science being a story of corrected mistakes, however, was not quite felicitous, because it describes the evolution of theories somewhat incompletely. In practice, scientists often refuse to discard their pet theories, and a great many excuses are invented to shield them against the pile-up of negative evidence before the once-fashionable model is eventually thrown overboard. This phenomenon pervades all of physics, and it is worthwhile to review a couple of these excuses by putting them into a broader context.

A LESSON FROM HISTORY: NO EVOLUTION WITHOUT REVOLUTION

Unlike Popper, the philosopher Thomas Kuhn focused on the resistance of physicists to letting go of their favored but wrong theories, which often hindered progress in science. It could be said that Popper was the moralist of science who formulated its rules, and Kuhn the realist who described how these very rules were being systematically undermined. For that reason, some feel much more inspired by Kuhn.

Kuhn correctly observed that there were periods of "normal science," which are punctuated with short, violent revolutions. During the normal periods, scientists gather data and interpret it according to current views, such as the geocentric worldview of Ptolemy. Eventually, so many anomalies are piled up that the old picture becomes confusingly complex and vulnerable to a paradigm shift.

In the Copernican revolution, the center of the world literally shifted from Earth to Sun. In the "normal" periods of science, poorly understood phenomena typically surface and are then integrated into the existing model with new, and often ad hoc, assumptions. That's why medieval astronomy had swelled into a junk room. And as Kuhn noted, there is no possible compromise between two "incommensurable" world views: if a model is wrong, it's not a little wrong, but completely wrong.

According to Lee Smolin, there are two types of scientists that correspond with the normal and revolutionary phases of science—the team-oriented "craftspeople" and the unconventional "seers."* String theorists tend to flatteringly categorize themselves as visionaries, seemingly unaware that a whole field of individualists is actually a monoculture.

By comparison, those who observe precisely are advancing the workmanship of science, including the numerous physicists who participate in the collaborations of particle physics or cosmology, albeit within sociological structures that suppress reflecting on basic problems. The task is to describe, not explain, which typically leads to a creeping complication. An obvious sign for this is the increase of free parameters, the unexplained numbers that Nature seems to be maliciously foisting upon us.

WHAT DOES INTERACTION REALLY MEAN? AND IS FOUR A HOLY NUMBER?

Complications in science may appear in many different ways, even as a new mechanism. For example, the common classification of four interactions—namely gravitation, electromagnetism, and weak and strong nuclear interaction—is indeed peculiar. Why four? Why are gravitational and nuclear forces attractive, while the electrical force is both attractive and repulsive? This difference isn't deeply understood by scientists at all, and it may well

---

* Other terms are "hill climbers" and "valley crossers."

be that we are lumping together very different things that have no more in common than fire, earth, water, and air, the four elements describing the universe of the Greek philosopher Aristotle.

Equally narrow is the idea that all four interactions are due to a particle exchange. The graviton, which is supposed to be responsible for gravity, has only been found in theoretical articles so far. And wasn't it quantum theory, with its wave-particle duality, that taught us that the microscopic world can be described neither by waves nor by particles on their own? This appears to have been long forgotten, triggered by the euphoria resulting from the flood of new particles in the postwar era. In 1950, Enrico Fermi made a legendary statement in *Physical Review*: "In recent years several new particles have been discovered which are currently assumed to be 'elementary,' that is, essentially structureless. The probability that all such particles should be really elementary becomes less and less as their number increases."[1]

Fermi's uneasiness evidently came from the complication caused by so many new particles. But today this number has grown even larger, and there is nothing preventing theorists from postulating new particles to "explain" new phenomena. The hypothetical inflation in the early universe, not surprisingly, postulates a particle called "inflaton." The "pomeron" and "preon" particles have been proposed as constituents of the atomic nucleus, and the "cosmon" might be responsible for some cosmological anomaly. Sure, why not. But this is a lot like thinking up names for a pantheon of Roman gods. I suppose that there could be a "theoron," which interacts solely with the brains of theorists, and might account for this blindness.

If stupidity got us into this mess, why can't it get us out?

—*Will Rogers, American humorist*

LIGHT DARK MATTER AND DARK DARK MATTER, AND WHATEVER MAY FOLLOW

Extremely well-shielded underground particle detectors such as the Gran Sasso laboratory in Italy show the huge efforts under way to detect the long-sought-after dark matter particle. The experiment is so sensitive that smokers are not allowed anywhere near it. Tiny amounts of radioactive substances contained in the fumes penetrate their clothes, and that risks contaminating measurements.

Despite all caution, distilling the desired signal from the shower of unwanted background signals is a severe problem. Recently, John Ralston, a nuclear physicist from the University of Kentucky, harshly criticized the careless methods of analyzing the detector data of the underground laboratories.[2] Neutrons originating from the cosmic-ray bombardment of Earth may well mimic dark matter particles. The problem is that generations of PhDs don't reflect upon how poorly their vastly simplified models describe these neutron reactions.

Almost inevitably, artifacts are produced, as in the case of the DAMA/LIBRA-experiment, a dark matter search in the Gran Sasso laboratory. The researchers found a summer-winter fluctuation in detector signals,[3] which has remained stubbornly unconfirmed by other researchers. The effect nonetheless generated a really neat interpretation. Since the velocity of Earth with respect to the rest of the Milky Way varies slightly due to the orbit around the Sun, the favored explanation was that our detectors hit the elusive dark matter particles at different speeds.[4] Of course, why not? The real problem behind this is that the scientists' expectations are biased even if they try to be objective, as Ralston comments:

"Nobody wants the field of dark matter detection to follow the slippery slope of unprogressive conservatism where higher and higher status will be achieved by developing better and better technology to discover nothing."

In this vein, we are facing another interesting interpretation of a recent experiment. PAMELA and AMS, two satellite experiments detecting energetic cosmic particles, found that with respect to electrons, there were too many positrons (the antimatter partners of the electron). Here again, one may cook up an "explanation" by postulating that the positrons could have been emitted while one sort of dark matter transformed into another!

For 80 years now, we have been chasing dark matter without success, and we don't have the slightest clue about what it could be made up of. But that hasn't stopped people from starting to classify it into subspecies. Naturally, other researchers are waiting in the wings to verify the "discovery."* Hypothesizing two invisible things will surely fit the data better than one. What a progress!

> Nobody ever won a Nobel for proving that something didn't exist or some theory was wrong.
>
> —*Gary Taubes, science author*

DARK SUBSTANCE OR DARK PHYSICS?

Another comfortable tool for tuning cosmological observations is the time evolution of the cosmological constant Lambda. Keep in mind that Lambda itself was a description of an anomalous time evolution of the universe—the

---

*Indeed, the astronomer Pavel Kroupa from the University of Bonn predicted that in the next couple of years *something* will be declared as dark matter.

seemingly accelerated expansion of the cosmos found in 1998. Above all, this is a poorly understood effect. Remedying with the parameter Lambda helped for about ten years, but again, the relentlessly increasing precision of measurements seems to indicate a discrepancy with the picture of the cosmos that we have barely digested. The next round of complications will lead to a temporal change in Lambda, or a dark matter–dark energy interaction,[5] or whatever.

But why should the acceleration of the universe's expansion change? Why is the Hubble expansion accelerating at all? And yet a more profound question, why the hell is the cosmos expanding? The more parameters we use for description, the more superficiality we introduce into our description of the universe, and the more the fundamental questions we ought to understand are swept under the rug.

It is a similar game when we declare matter to be "dark." At the base of it is the fact that we just don't understand the high galaxy speeds in clusters and the too rapidly rotating galaxy edges. Whether you like to call it dark matter, or you prefer to introduce a parameter such as acceleration a_0 in the MOND theory, both approaches are epistemological complications. Maybe dark matter and its adjustable distribution are still more flexible than a single value a_0. But this is not the big advantage as cosmologists see it. There is too much dark stuff we have borrowed with assumptions. One day observations must pay off.

The too-marginal temperature fluctuations of the cosmic microwave background were declared as independent evidence for dark matter, but in the end, only a new free parameter smuggled into the model could account for the contradictory galaxy-formation process. The recent history of cosmology is a continuous production of such unexplained numbers. It is always much easier to resort to such patchwork than to search for explanations based on first principles.

NEUTRINO OSCILLATIONS: FROM EMBARRASSING DISCREPANCY TO CONFERENCE TOPIC

In the history of particle physics, one comes across a very similar mentality of doing research. Neutrinos, a decade-long candidate for dark matter by the way, failed to be registered on Earth in a sufficient quantity to correspond to the nuclear fusion rate inside the Sun. Luckily, theorists had provided three kinds of neutrinos, all of them verified with great effort by experiments that fewer neutrino species could not explain.

The more particles you have, however, the more phenomena you can explain. The seemingly missing neutrinos from the Sun could now have just transformed into either of the two sister particles, which are invisible to the detector. Since then, conferences have been organized about these "neutrino oscillations" between sister particles, and the problem has been transformed into a discovery.

Modern observations in the Japanese Kamiokande detector and at the Sudbury Neutrino Observatory in Canada claim to support neutrino oscillations, though measurements usually show visible neutrinos disappearing rather than invisible ones reappearing. Whatever the correct interpretation might be, it is impossible to deny the diagnosis of a severe complication— three neutrinos with different masses, reaction probabilities, and tendencies to transform into each other, called "mixing angles." As usual, a couple more numbers complicating the description, and no theory in sight capable of explaining them.

PHYSICAL PHANTOM PAIN—THE BIRTH OF NEUTRINOS

> Don't be trapped by dogma—which is living with the results of other people's thinking.
>
> —*Steve Jobs*

We might even consider the heresy that the entire idea of neutrinos itself originated out of a lack of understanding. Around 1930, when the decay of a neutron into a proton and electron was put to the test, it seemed to violate Einstein's beautiful energy formula $E = mc^2$. On average, half of the energy the electron was supposed to carry was missing! No one could come up with a satisfying explanation. Wolfgang Pauli eventually called in the neutrino, assigning it the amount of missing energy. Subsequent experimental observations vindicated him, although it took 25 years for an elaborate experiment to be conducted.[6] The cross-section of a neutrino is so incredibly tiny that every second, billions of neutrinos pass through our bodies without even a single one being absorbed. Nobody knows why Nature produces such elusive particles.

There is also a mystery about why no photon is involved in the decay of a neutron into a proton. In all other contexts in physics, particles that are accelerated emit electromagnetic radiation. Although the sudden emission of the electron during a neutron decay must be accompanied by a huge acceleration, no photon comes up, whether we like it or not.*

Around 1930, however, new particles were considered as unpleasant complications. Wolfgang Pauli reluctantly put forward his neutrino idea,

---

* Interestingly in a small percentage of recent cases this was indeed discovered (J. Nico et al. Nature 444 (2006):1059ff.). This, however, doesn't solve the riddle as a whole.

and Enrico Fermi allegedly joked that he would give him a bastinado for having put the blame of the missing energy on an innocent new particle. Pauli's proposal was seen as a trite excuse by everybody, and even seven years later, the Nobelist Isidoor Isaac Rabi commented on the discovery of the muon with irritation, demanding to know, "Who ordered that?"

Today, new particles are not regarded as methodical defeat anymore, nor an oath of bankruptcy as in the 1920s. Just as a gambler does not bother with unpaid bills and IOUs, we don't care about a healthy frugality in the laws of Nature, but rather continue to cheerfully go shopping with our accelerators.

ANTIMATTER, THE UNFORTUNATE TWIN

No matter how much energy we use in our colliders to smash particles, the collisions always result in the creation of the same amount of matter and antimatter. But the universe must contain far more matter than antimatter, otherwise astronomers would long ago have detected the battlefields marked by the radiation of matter and antimatter particles annihilating each other. This matter asymmetry conveys a mystery. It is assumed that right after the Big Bang, a little more normal matter than antimatter was left by coincidence. A tiny fraction of about one particle of matter to a billion of particle-antiparticle pairs seems to be consistent with observation. Of course this is not an explanation but rather another free parameter put into the model. Nobody can explain why the above ratio is 1 to 1 billion and not 1,000 or even 1 trillion. No one knows. And we haven't the slightest clue how to calculate this number, as the matter/antimatter particles behave identically in the laboratory.*

COLLIDERS' CAPERS

Nothing is created by coincidence, rather there is reason and necessity for everything.

—Leukippus, ancient philosopher

After a long period of theorizing and interpreting complex experiments, physicists convinced themselves that the parts of the nucleus—the proton and neutron—were made up of other parts themselves, the so-called quarks. The particle physicist and science historian Andrew Pickering called this "a social symbiosis of experimental and theoretical practice"

---

* It is claimed that there is a tiny discrepancy while observing the kaon decay.

in his excellent book *Constructing Quarks*. The initially simple world of *up*- and *down*-quarks was, step-by-step, extended, and two kinds of quarks turned into six. For not running into contradictions, this was followed by tripling the number of quarks by introducing "colors," epistemologically one of the most dramatic complications of the standard model.

Evidence for the first quarks was seen in the fact that protons did not behave like homogeneously charged spheres when hit by energetic electrons, and thus should have a substructure of some kind. But is this a real proof for quarks? Rather, we should keep in mind that we still don't understand such processes involving highly accelerated charges.

And we don't understand either why particles, having a spin, don't obey that spherical symmetry. Even the otherwise so profoundly arguing Richard Feynman creates the impression that Nature did some superfluous gimmickry while inventing the spin. The calculations would be so much easier without it.[7] However, a reason for spin must exist. And there is yet another very weird property of the spin. The proposed three quarks that make up the proton are themselves imbued with a spin that may point in whatever direction. Why they align and neatly add up to the proton's spin value is mysterious. This problem, called the "spin crisis," shows that there is a deep-rooted lack of understanding inherent to the very concept of spin.[8] Again, it seems that the inexpensive "solution" of one problem by postulating quarks has produced a more subtle one.*

How long will this go on? Continually subdividing elementary particles into smaller and smaller building blocks has reached a momentum that we should call into question at some point. Or should we just wait for the "discovery" of quark components?

DISTINGUISHING THE DIFFERENCES THROUGH UNEQUAL EQUALITY

They had to double, in a stroke, the number of elementary particles in the world, and then declare, that, by mischance, half of them had not yet been found.[9]

—David Lindley

The still-mysterious classification of elementary particles is the reason behind proposals such as supersymmetry, which complicate things even more dramatically. Instead of adding arbitrary numbers one by one as needed, as in the "conventional" erosion of science, in hope of admittance

---

* The methodical parallel in cosmology is the coincidence problem that came up after resolving the cosmic age problem with dark energy.

to the Promised Land of the grand unification of physics, the entire set of parameters is doubled. Some adherents of supersymmetry have claimed that their idea is analogous to Paul Dirac's prediction of antiparticles. This is grossly misleading, since back then, not a single parameter was added. All antiparticles have the same mass as their partners. The particle clones of supersymmetry, on the other hand, are supposed to be a lot heavier than their partners for some unknown reason. The fashionable explanation is "symmetry breaking," a vague term physicists use as an easy excuse for just about anything. If you ask for a deeper meaning, you catch them on the wrong foot. (This would also qualify as "symmetry breaking.")

The Nobel committee was not beset by doubts about whether or not symmetry breaking is a fundamental mechanism of particles, because in 2008 it awarded a prize for its discovery. Symmetry breaking as such is surely an interesting phenomenon, occurring, for example, when water transforms from the aqueous to the gaseous phase. An evenly heated pot of boiling water releases its steam bubbles asymmetrically. But these "explanations" are getting out of hand. The bubbles become babble.

> The rule for explaining every seeming distinction: turn it into a perfect symmetry, improperly realized.[10]

> —David Lindley

THE EVOLUTION OF THE UNIVERSE: THE EARLIER, THE MORE COMPLICATED?

In cosmology, our lack of understanding culminates at the Big Bang. Because symmetry breaking does such a great job in explaining the unexplainable, it has become the bread and butter of inflation theory, which deals with the first moments of the cosmos. But inflation was an excuse invented because we don't understand the "flatness" of the universe, that seemingly fine-tuned state on the edge between eternal expansion and future contraction.

Cosmologists are looking at the first billion years after inflation, and are again fiddling around with an ad hoc mechanism. Do the large voids in galaxy structures contradict the models? No problem. Fix it with the assumption of a sudden divorce of all electrons and atomic nuclei called reionization, which "explains" why we don't see what we should see. Whether you believe in reionization or not, it doesn't exactly simplify the description of the early days of the universe, not to mention the extra parameters used for reionization to patch up the model. Generally, the farther we go back in time, the more complicated the physics. Not really a reassuring scenario.

WHY THE PILING-UP OF CONSTANTS SHOULD MAKE YOU LEERY

The sum of obvious little steps is not seldom a way in the wrong direction.

—*Richard David Precht, contemporary philosopher*

Many practitioners of physics will not share my doubts, and they will come up with evidence that makes every single complication appear reasonable. But even if a model manages to be justified by the data, it is evident that the big picture has gotten too messy. Some key concepts have developed subtly over generations of researchers, so they don't stand out in the daily science routine. But the day will come when we will have to take a closer look at the balance sheet, and realize that there are many, many numbers that are not understood.

The classic example of the erosion of scientific constructions is the geocentric picture of medieval astronomy, with its many epicycles that prevailed for 1,500 years. But there are also much shorter episodes that should make us reflect. In the 1930s, shortly after Enrico Fermi had discovered that he could induce nuclear reactions with slow neutrons, he split the first nuclei. Fermi remained unaware of his discovery because it contradicted the prevailing dogma that splitting up the nucleus required high energies. In 1934, the chemist Ida Noddack was mocked for her suggestion that the nucleus could be split.

However, researchers could not make sense out of the numerous fission products. Thus they were falsely classified as very heavy nuclei called "transuranic elements," upon which a complicated theory was built. Ironically, Enrico Fermi was honored with the Nobel Prize for this misinterpretation. In 1938, Otto Hahn and Fritz Strassmann in Berlin, after stubbornly having defended the wrong model, eventually were confronted with inescapable evidence for nuclear fission.

Five years of research were lost. This doesn't seem like much, but compared to the unsolved problems of modern particle and astrophysics, the atomic nucleus was a tiny mystery. We must become aware of how easily science ends up on the wrong track.

DO WE NOTICE RIGHT AWAY IF SOMETHING GOES WRONG?

Once the error is placed in the ground like a foundation stone, everything is built on it, nevermore it turns to light.

—*Rudolf Clausius, nineteenth-century physicist*

When I talk to people about general relativity and its tiny effect on time lapse at high altitudes, I like to mention that GPS satellites wouldn't work properly without Einstein's genius. But is that really true? If atomic clocks in satellites had been built without the theory of general relativity, people might then have stumbled upon the fact that time measurements seem to depend on the altitude and velocity. To adjust this with two parameters there is surely no need for an Einstein. And I fear that the observation of one parameter being in the order of magnitude of GM/rc^2 —a consequence Einstein had deduced—would have been disregarded as a mere playing with numbers, or "numerology." Looking at the history of science, it may easily happen that we prematurely give up the search for explanations by fundamental theories and instead cement our incomprehension with a patchwork model.

Senior scientists often greeted the model extensions with healthy skepticism, for the very reason that these models did not come with convincing explanations. But the slower science progressed, the more time remained for the skeptics to die out. The new concepts weren't accepted enthusiastically, but rather with creeping resignation and a "we-don't-have-anything-better" attitude. Statements such as "we are going to establish the standard model of cosmology just as firmly as that of elementary particles"[11] should give you cold chills, as if *that* messy model was a model mission to follow.

The painstakingly constructed model of particle physics does not possess any attraction either, if I may resort to the argument of aesthetics for once. In my daily contact with young, still-innocent people, I see how much fascination physics of the years up to 1930 elicits, while its modern extensions are met with indifferent faces. Students who had, up to that point, preferred physics over biology and chemistry are then disappointed. The standard models of particle physics and cosmology can—notwithstanding the utter lack of understanding—be memorized, while the principles of the theory of relativity and quantum mechanics can—notwithstanding the difficulties—be actually *understood*. This is another feature distinguishing good physics from the cheap descriptions.

> Instead of filling a gap by guesswork, genuine science prefers to put up with it.
>
> —*Erwin Schrödinger*

GRAFFITI ON THE WALLS OF SCIENCE

Comparing the inflationary growth of "laws" in the standard models of particle physics and cosmology to the theorems of classical physics resembles the difference between detailed legislation gotten out of hand and the few but important articles of the US Constitution. This otherwise very superficial parallel between juristic laws and those of Nature can only indicate that the latter ones are, at the moment, made by humans, too! The rapid growth of complexity is a sign of the erosion of good science, but at the same time, that very complexity is making it more difficult to identify the flaws. In view of his colleagues' attitude, the cosmologist Fred Hoyle once commented, "They defend the old theories by complicating things to the point of incomprehensibility." But a disastrous co-evolution has especially developed in particle physics. Many observers are happy to squeeze new data into the framework of their models. True research is replaced by statistical kneading, which is supposed to legitimize the many parameters. And if despite all efforts, new contradictions show up, the reaction is as follows:

If physicists do not understand the *what* of their theories, they'll introduce a new particle. If they don't understand the *when,* then it must have happened right after the Big Bang. If they don't understand the *where,* then of course it took place in an extra dimension. And if they don't understand the *how,* they will postulate a new interaction. If they don't understand the *how much,* a symmetry breaking will soon appear. If they don't understand anything, they will propose strings and branes. And if they lose interest in all understanding, there is always the strong anthropic principle. Things have come to a pretty pass.

All the couplings, particles, densities, fields, fluctuations, and parameters are nothing other than what Richard Feynman termed as crap piled on old theories. Isaac Newton's words cannot be repeated often enough—truth lies in simplicity. One cannot trust this complicated structure anymore. The crash is coming.

Experimental Physics:
A Reality Check

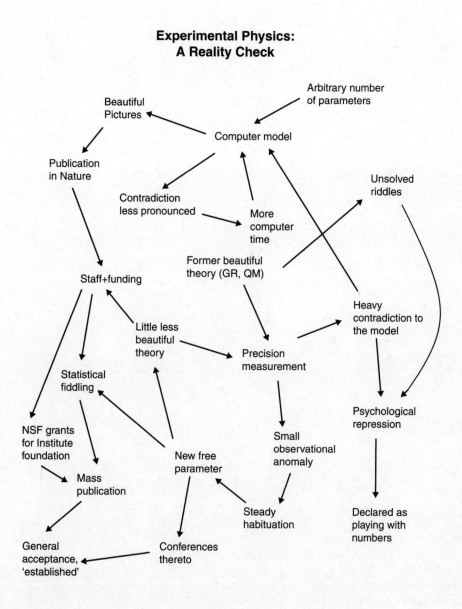

Chapter 18

THE FIRST WRONG TURN

DEVIATION DECADES AGO: CALCULATING REPLACES THINKING

I f my skepticism in the last chapter concerning the established concepts of theoretical physics got on your nerves, then you may now need to take a Valium. We really only have two possibilities: either we are content with the notion that Nature is a big mess of confusing structures that we are unable to explain—or we scrutinize its laws in the pursuit of simplicity by challenging the free parameters.

There are more than 50 unexplained numbers in the standard models of physics, which means that physics is in a desperate state. But physicists readily bring forward good reasons to ignore the demand for simplification. Therefore, I will stand out as a hopeless optimist with my confidence that one day theoretical physics should manage without *any* free parameters. Evidently, this dream leads back to a time long ago, when physics rested its modest number of theories upon a few fundamental constants. Let's have a closer look at these mysterious quantities.

> In a reasonable theory, there are no numbers which can be only determined empirically.[1]
>
> —*Albert Einstein*

QUANTUM ELECTRODYNAMICS—HAVING SAID A BIT TOO MUCH

Quantum electrodynamics, developed in the 1940s to describe what particles are made of, is a good theory for the very reason that it allows precise predictions without introducing any new parameters. The theory is further reinforced by its ability to calculate a tiny discrepancy of the wavelengths in a hydrogen atom using the well-known fine structure constant of about 1/137. That is a great result.

But there are some problems, too. Although the name "quantum electrodynamics" suggests that this theory is a satisfying synthesis of electrodynamics and quantum theory, it is not. Otherwise, it would be possible to calculate the fine structure constant from first principles. If we could derive this number, it would allow the electrical constants to be expressed through Planck's constant! Thus, the number of fundamental constants—which are free parameters as well—would be reduced by one.

Not only Richard Feynman, but also Paul Dirac, considered this to be of outstanding importance, such that whenever a theorist wanted to present a new idea to him, he would demand first, "Can you calculate the fine structure constant with this? No? Come again, when you are able to."

There is, however, another deficiency of quantum electrodynamics, which Feynman himself had often pointed out. If we calculate the electron's energy using the formula for its electric field, it results in an infinite value. Remember that Einstein's $E = mc^2$ (the energy-mass equivalence) would turn that into an infinitely heavy electron, which is absolutely nonsensical. Quantum electrodynamics dodges this problem by assuming a "naked" electron, one without an electrical field and an infinite amount of negative energy, whatever that is. Then, abracadabra, the naked electron, dressed with the electric field, matches the observations. Although calculations based on this trick for removing annoying infinities, called renormalization, have had successful results, Feynman remained doubtful about its validity:

"It's surprising that the theory still hasn't been proved self-consistent one way or the other by now; I suspect that renormalization is not mathematically legitimate. What is certain that we do not have a good mathematical way to describe quantum electrodynamics."

The ever-skeptical Dirac made his opinion of renormalization quite clear: "Sensible mathematics involves neglecting a quantity when it turns out to be small—not neglecting it just because it is infinitely great and you do not want it!"

Nonetheless, the entire theorizing in particle physics,* named "quantum field theory," is based upon this concept. To get a picture of how deeply renormalization is embedded in the theory, you may want to look up the official Nobel Prize for Physics announcement of 1999. The word "renormalization" was used 24 times.

---

* There is only one field that stubbornly resists renormalization: gravity. The reason for this, as for many other mysteries, lies in the units of G.

Once again, productive computing has prevailed over fundamental reflections. The infinities, as well as the origin of the fine structure constant, remain unsolved puzzles, which continue to fade from our attention.

> In order to progress, we must recognize our ignorance and leave room for doubt.
>
> —*Richard Feynman*

IS SOMETHING WRONG WITH THE GRAVITATIONAL CONSTANT AFTER ALL?

Einstein's theory of general relativity is excellently tested, similarly precise as quantum electrodynamics. Its predictions perfectly agree with the so-called classical tests, such as the bending of light due to the gravitational effect of the Sun, that were discussed in chapter 4. As in quantum electrodynamics, this is achieved without a new free parameter. The small corrections to Newton's theory contain just the term $2GM/rc^2$ with the already known constants G and c. Unlike quantum electrodynamics, general relativity does not suffer from inherent inconsistencies, but neither does it reduce the number of fundamental constants either.

The all-encompassing gravitational constant G is implicated in a whole series of problems in physics, such as the amazing properties of black holes, the actual sizes of which have never been measured. Black holes are widely considered a consequence of the theory of general relativity, although Einstein himself did not trust the mathematical extrapolation to such extreme cases.

The very tiny Planck length, a direct consequence of the gravitational constant, has become the playground for unsuccessful attempts at describing quantum gravity. This has produced a really wide gateway into hokum with the "holographic principle," according to which all information in the universe is encoded in its border. Now, just try to go out there to the border to see!

Due to the unchallenged status of general relativity in the physics community, barely anybody questions whether G is a true constant, although Einstein admittedly left Ernst Mach at the door, the Viennese physicist and great thinker who had inspired him.

FROM NEWTON, PASSING BY MACH, TO EINSTEIN

It was the British-Egyptian cosmologist Dennis Sciama who took up Mach's ideas again. In 1953, Sciama wrote, "Newton's theory contains two arbitrary elements, namely the choice of the absolute reference system and the value of the gravitational constant G."*

Removing the arbitrariness of G, however, would mean being able to calculate it from first principles. This would reduce one of Nature's big free parameters and revolutionize physics, just as would the computation of the fine structure constant. The rate of expansion of the universe and its mass would no longer be unexplained numerical values, but rather reflect the laws of motion and the gravitational constant—an intriguing scenario!

Sciama tried this. He simply considered the well-known gravitational potential of a body with mass M at a distance r, which is GM/r, and used it to calculate the gravitational potential of the entire universe.[2] The value he got had the order of magnitude of c^2, the square of the speed of light. Despite the uncertainty resulting from estimating these quantities, this is an incredible coincidence, given the very large numbers being dealt with. Sciama was the first to find a formula reflecting Ernst Mach's conviction that inertia (mass) is related in a fundamental way to the presence of all other masses in the universe.**

However, Sciama could not develop a consistent theory out of his hypothesis. His formula did suggest that G was a variable and not a constant, but without experimental indications to support it, the idea went back to sleep. It's not that the time variation of G as such would be fascinating, but rather the possibility to principally compute G using other data.

Maybe the mystery of the origin of mass, which pervades all of physics, has its roots precisely here. In any case, approaches like Sciama's, even if speculative to date, are the only conceivable road at this point to reducing the number of free parameters in the theory of gravitation. G would be the perfect candidate for it.

CONTRADICTIONS IN ELECTRODYNAMICS—SQUEEZED OUT, NOT SOLVED

There are infinities that arise when applying the general theory of relativity to black holes, just as there are infinities in quantum electrodynamics. But infinities, as we have seen, are just thrust aside by invoking renormalization. Richard Feynman and Lev Landau, the Russian theoretician, have independently said that this makes it impossible to calculate the radiated energy of a strongly accelerated charge,[3] exactly the kind of thing physicists

---

*The first-mentioned problem of "Newton's bucket" was solved by Donald Lynden-Bell and Jonathan Katz in a wonderful, but widely unknown paper (arXiv:astro-ph/9509158).

** A little later, the American astrophysicist Robert Dicke developed similar thoughts. I wrote a slightly more technical paper about Dicke's and Sciama's approach (arXiv.org/abs/0708.3518).

would need to know when firing up particle accelerators and smashing particles in them.

How do physicists conducting the entire business of collider experiments come to terms with this blind spot in our knowledge? That remains their secret. It seems that generations of particle physicists have distracted themselves with patchwork formulas and relegated the problem to the subconscious.

Once you realize that any acceleration of a charge leads to radiation losses, then Newton's second law of motion, $F = ma$ (force equals mass times acceleration), is not precisely valid any more. Possibly, the very nature of "forces" in electrodynamics is very different from the mechanical forces. It may just be that their almost similar behavior leads us to lump them into one term.

RELATIVITY, ETHER, AND THE WAVE AS A PARTICLE

Notably, mathematically simple equations (so-called linear equations) describe electrodynamics, and the contradictions arising from this pose the question of whether this is *too* simple. Throughout physics, a truly realistic description often requires intricate nonlinear equations, as in fluid mechanics or continuum mechanics (which deals with elastic substances such as rubber). Incidentally, for the entire nineteenth century, continuum mechanics was believed to be a valuable description of electrodynamics. Physicists imagined electromagnetic waves as propagating oscillations of an elastic medium called the ether, which was believed to permeate all of space.

The ether theories vanished after 1905, because Einstein's theory of relativity didn't need ether and the experimenters at that time couldn't find it. In fact, the idea of masses gliding through the ether as fish do through water leads to contradictions.

However, there is an intriguing analogy. Wave structures and other irregularities in an elastic continuum (defects) surprisingly behave like particles. Being nothing but the nucleus of a disturbance, it cannot move faster than the disturbance itself, and the motion of the disturbance is limited by the speed of sound. The analogy to electron motion, which is limited by the speed of light, is obvious! Moreover, the formulas for charged particles in the special theory of relativity are identical to those describing the motion of defects in elastic solids.[4] This is exciting, because it could mean that the ether was abandoned prematurely, since people didn't know about the possibility of modeling particles as defects in such an elastic solid.

Despite all these open questions about electrodynamics, it has a great, everlasting legacy. By proving that light was an electromagnetic wave, the famous equation containing the electric and magnetic constants ($\varepsilon_0 \mu_0 = 1/c^2$) reduces three fundamental constants to two.* Here is the groundbreaking result of Maxwell's electrodynamics—one less free parameter.

THE RADICAL SIMPLIFICATION THROUGH h

The most important pillar of modern physics is quantum mechanics. Planck's radiation law of 1900, with its famous quantum h, encompassed the formulas of earlier theories about radiation and rendered their constants unnecessary. The birth of quantum mechanics immediately reduced the number of free parameters. And in all the revolutionary insights that followed, such as Einstein's light quanta and Louis de Broglie's wavelength of the electron, h played an essential role of simplification.

The chain of discoveries involving h continued, up to the Nobel Prizes in 1972 and 1985 for subtle microscopic effects. However the absolute highlight was Niels Bohr's insight in 1913 that h took the role of an angular momentum in an atom. It led right away to the calculation of the mysterious number that Johann Jakob Balmer had extracted from the atomic spectra in 1885. That number—a free parameter—was explained satisfactorily and went away.

QUANTUM MECHANICS—THE BEST NOT UNDERSTOOD THEORY

It seems clear that the present quantum mechanics is not in its final form.

—*Paul Dirac*

Quantum mechanics erased free parameters in many instances, and no other theory appears to have done a better job of simplifying the laws of Nature. Yet, more than any other theory, it has also made the laws of Nature appear to be incomprehensible.

For the first time in the history of human thinking, randomness seemed to play a role in describing the universe. For many it was unacceptable that Nature's decisions should not be computable on the microscopic scale. Specifically, Schrödinger and Einstein objected, the latter allegedly commenting with the pithy words "God doesn't play dice" (which was not intended to be a religious statement, by the way).**

---

* The coincidence was first noted by Wilhelm Weber, a pioneer of electrodynamics.

** Einstein however, was proven wrong in his related claim that this randomness would forbid two particles to adjust their spins at a distance. Experimenters proved the "nonlocality" of quantum mechanics in the 1980s—one of the most puzzling facts ever detected.

A pragmatic but probably not fully satisfying interpretation was given by Max Born and later publicized by Heisenberg, Pauli, and Bohr as the "Copenhagen Interpretation." In this interpretation of quantum physics, the wave nature of particles (wave function) is not expressed as having any material meaning, but rather provides only the probabilities of finding a particle in a particular place and state.

A series of fundamental questions arise here, such as how it is possible that math with rules more suited to gambling suddenly turns into something of the real physical world. Such questions have been debated at considerable length by physicists. Bohr and Born, in particular, went to great lengths trying to give their findings a meaning outside the realm of physics.[5] Some exhausting discussions took place without leading to any concrete results.

KEEPING AWAY FROM INFECTIONS OF NONSCIENCE

Purely conceptual arguments can be awfully strenuous. I attended a lecture at a course entitled "Advances in Quantum Mechanics" long ago in Erice, Sicily. Just before the lunch break, a speaker was long-windedly outlining the concept of a "protected" wave function and a "leakage" from it, although most of the audience did not understand what he wanted to say. Many growling stomachs protested against the igniting debate, when Michael Berry (who had become a luminary because of his discovery of a geometric relation in quantum mechanics) raised his hand. The assembly dissolved in laughter when he said, "The wave function is protected? Well I guess you have then invented the quantum condom. It distinguishes itself from the classical one in this aspect. It only protects when it leaks!"

Discussions of this sort may have led pragmatists such as Feynman to not bother racking their brains about how a mathematical abstraction like a probability function suddenly materializes into a real particle (the "collapse of the wave function"). This is understandable. The rules, seen as cooking recipes, just work well in predicting the observations. Imagination, the former guide to the greatest accomplishments in physics, surrendered here, giving rise to a "shut up and calculate" school of nonthinking. It is a shortcoming of quantum mechanics that the wave and particle behavior of Nature is not a logical consequence of the theory.

Maybe giving up the demand for an intuitive understanding of the underlying processes in the microscopic world early in the twentieth century was where the troubles with theoretical physics became entrenched. It is disconcerting that almost all founders of quantum mechanics had their own interpretation of it in mind. And from the beginning of quantum physics, a certain dogmatism over the predominant probabilistic interpretation was established, while other approaches, such as the central idea in the doctoral thesis of the Nobelist Louis de Broglie, fell into oblivion (sadly, nobody reads the original).[6]

Very few physicists today worry about the fundamentals of quantum mechanics. Whatever the progress might be, it must predict new experimental effects. Attempts at trying to reform quantum theory on a conceptual level only are in danger of drifting off into chitchat.

WHY DOES THE SPIN EXIST? WHERE DOES THE MASS COME FROM?

In 1928, Paul Dirac applied Einstein's theory of special relativity to quantum mechanics, ingeniously deducing a mathematical structure that described "spin"—the mysterious property of elementary particles that is a sort of rotation, but not quite. But why doesn't Nature allow for plain, boring spheres for its particles? Only a deeper understanding of quantum mechanics may be able to explain spin as a real consequence of the theory.

In its current form, quantum mechanics is also incapable of calculating the basic characteristics of elementary particles. Take, for example, the mass ratio of a proton and an electron, which is 1836.15. It eludes any comprehension. And there is no reason why the typical masses themselves lie in the observed order of magnitude, at least according to the common opinion. But there is a hint.

Planck's constant h approximately matches the product of the speed of light, the proton's mass m_p and its radius r_p, such that h is about equal to ($\approx$) $c\, m_p r_p$. The approximate value for the proton radius ($r_p = 1.3 \times 10^{-15}$ meters) has been known experimentally since 1914. So in some sense, the proton is the most typical "quantum mechanical" particle. Furthermore, it is stable, in contrast to the collider-generated zoo of particles to which we attribute so much significance despite the tiny life span of the particles.

There is reason to suspect that Nature wants to tell us something with the coincidence of h being so close in value to $c\, m_p r_p$. However, no theory exists to provide an explanation! If there were such a theory, we would be able to calculate one fundamental constant using the others, resulting in one less free parameter. We would take a leap toward solving the mystery of mass.

SIZE AND MASS OF THE PROTON: JUST A COINCIDENCE OR A MESSAGE FROM NATURE?

It does seem worthwhile to contemplate the relationship $h \approx cm_p r_p$. Surprisingly, it turns out to be equivalent to Dirac's hypothesis[7] about the number of elementary particles in the universe, which was already touched upon in chapter 14 (if you should be interested in a little more detail, see the hints in the appendix—otherwise just trust the calculation!). Dirac suspected that the ratio of electrical to gravitational force, approximately 10^{40}, was linked to the ratio of the size of the universe to the proton, equally about 10^{40}. Moreover, he noted that about 10^{80} (the square of 10^{40}) protons exist in the universe. The intriguing coincidence $h \approx cm_p r_p$ is therefore just another form of Dirac's conjecture.[8]

At the same time, Dirac's idea relates to the formula put forward by Dennis Sciama, who tried to understand the gravitational constant G with Mach's principle. But Dirac goes further. Sciama's idea alone, $G \approx c^2 R_u / M_u$, could be realized with a different number and different masses of elementary particles. If Nature had created, for instance, 1 trillion times more particles, each 1 trillion times lighter, that's still fine for Sciama's formula. But only Dirac's observation gives a hint why the particles are as many and as heavy as they are.*

DIRAC AGAINST THE REST OF THE WORLD

The riddles of God are more satisfying than the solutions of man.

—*G.K. Chesterton*

I should not conceal that Dirac's conjecture—that the number of particles in the universe (10^{80}) is related to its size (which is 10^{40} proton radii)—clashes with all established concepts. According to the standard model, in the very distant future, when the universe has expanded to 10^{50} times the proton size, there would still be about 10^{80} visible particles, not 10^{100}, the square of 10^{50}. If one believes conventional cosmology, Dirac's observations are pure coincidences that take the opportunity to fool today's astronomers.

And Dirac's conjecture triggers horror among particle physicists because it suggests that the quantum effects of gravity start at the size of an atomic nucleus and not at the far smaller, unobserved Planck's length

---

* The Nobel laureate Frank Wilczek, albeit obviously unaware of what has been written earlier on the subject, wondered about this in arXiv:hep-ph/0201222v2.

of 10^{-35} meters.* Though we certainly do not understand the nature of the officially responsible nuclear forces very well, its experts would hardly share Dirac's view.

Was Dirac a dreamer who did not grasp modern physics anymore, making a fool of himself with unfounded numerology? If one credits the standard models of physics, then yes. But if we see the necessity of using fewer free parameters instead, we have to take Dirac's idea seriously. Actually, it is the only idea since then that has the potential to explain mass. This would, however, mean that since about 1930, the last two or three generations of theoretical physicists have been working in a completely wrong direction.

SPECULATIVE OBSERVATIONS ARE NOT THEORETICAL SPECULATIONS

> Without speculation there is no good and original observation.
>
> —*Charles Darwin*

Although this doesn't portray anything new with respect to the above, Dirac's second hypothesis about the particle number can also be illustrated as follows. If you lay out all protons in the universe as a carpet that is one proton deep, you will get a surface that covers the observed horizon of the universe. Somewhat surprising, isn't it? Numerical coincidences of this sort can rightly be called speculative, but it would be a great disservice to science to throw this into the same pot with theoretical speculations such as extra dimensions. Today, we can find numerous fantasies of technically well-engineered theories that can be tested in the distant future only, if ever. Here one may be well advised to wait for supporting observations before building too much upon these theories. In contrast, speculations of the sort Dirac made do not yet have a worked-out mathematical frame, although the observations are apparently there. But the majority of physicists are adhering to a statement from Sir Arthur Eddington (which he actually meant ironically) "not to put overmuch confidence in the observational results...until they are confirmed by theory."

ROCKING THE FOUNDATIONS AND THE ANGST OF DOING SO

Among the fundamental theories of physics, the theory of special relativity is the least controversial. This may be the reason why its basic principle,

---

* That this is 20 powers of ten below the nucleus' radius is a result of Dirac's hypotheses. However, it doesn't have any fundamental meaning.

the constancy of the speed of light c, in a sort of anticipatory obedience, is taken much more seriously than intended by Einstein.

The speed of light does not change if one looks at it from a moving car. That's relativity. Nevertheless, there could be a *spatial* variation of the speed of light, since all you need to do is change time and length scales of the things we measure in such a way that we do not notice a difference.[9] At first glance this sounds like a useless complication, but it can be used to reformulate the *general* theory relativity. A varying speed of light causes light to travel on curved paths, just as is the case of a lens bending the light ray to its focus. Almost every textbook mentions this point of view,[10] and the first attempt of this sort was undertaken in 1911, by Einstein. In 1957, the American astrophysicist Robert Dicke discovered a relationship between Mach's principle[11] and a variable speed of light, one of my pet themes I can just mention briefly here.

In chapter 10, we saw that some folks are more Catholic than the pope when they claim that a variable speed of light would be in contradiction to the theory of special relativity. But this belief is just in contradiction of Einstein, who wrote in 1911, "The constancy of the speed of light can only be maintained for space-time regions with constant gravitational potential."[12]

IF WE WANT TO UNDERSTAND SOMETHING, IT HAS TO BE VARIABLE

Most people prefer to speculate about difficult things rather than finding the truth in the simple ones.

—*René Descartes, French philosopher*

The belief in unalterable fundamental constants has become almost a dogma in the physics community, bringing to mind the medieval belief in an eternal starry sky. In order to achieve a true understanding, we often have to allow for the fact that changes in the universe may be taking place so slowly that we cannot see them. The gravitational constant has some suspicious properties, and the unexplained acceleration of the universe may mean we are overlooking a cosmological variability in the laws of Nature.

But above all, we can only work toward reducing the number of free parameters used to describe the universe if we strip away the holy constancy of these fundamental "constants." Either you take it as gospel that the mass and density of the cosmos and its particles—indeed, the whole universe—came into being as the result of arbitrary initial conditions. Or, you believe that this data should have something to say about Nature's laws. Figure out which way you want to go.

Before achieving the ambitious goal of eliminating free parameters, we may have to redefine the terms "space," "time," and "mass," and comprehend them in a deeper sense. It is naïve, for instance, to imagine a ticking time that is independent of cosmological evolution. Mach and Dirac, but also Barbour, thought one step further. A temporal change of the laws of Nature or its constants would of course have the greatest impact in the early universe.

A truly satisfying theory of the universe would likely be one where we have eliminated all these arbitrary numbers. Whether Nature ultimately allows such simple laws is not clear, but I feel uncomfortable that just about everybody else seems to have taken the opposite for granted. Einstein said:

> I'd like to state a law of Nature which is based upon nothing more than the belief in simplicity, that means comprehensibility of Nature. There are no arbitrary constants...that is to say, Nature is so constructed that it is possible logically to lay down such strongly determined laws which only contain logically deduced constants."[13]

We are far, far away from this becoming true.

Chapter 19

THE MATH FALLOUT

HOW THEORETICAL FASHIONS
IMPEDE REFLECTION

In the late 1800s, David Hilbert, Felix Klein, and Hermann Minkowski, all great mathematicians at Göttingen University, were advancing the idea that new mathematics could be applied to the unsolved problems in physics. Indeed, the early twentieth century has seen the most fruitful interplay between math and physics. However, of the fundamental problems that were still pending in 1930, none have been resolved by advances in mathematical physics. And alarmingly, in recent times, mathematical methods for describing Nature are being developed without bothering with experiments. Today's predominant view in the physics community, that new mathematics will eventually get physics out from its dead end, is wishful thinking.

Of course, theoretical physics needs some mathematical tools. Vector analysis describes fields such as the electric or magnetic ones. Complex numbers and differential equations in quantum mechanics help, for example, to describe the hydrogen atom. And to discover such differential equations, there is a technique called "calculus of variation." Finally, a thorough analysis of space-time in the general theory of relativity requires a bit of differential geometry. But more or less, that's all the math we need to deal with the three-dimensional world we perceive.

Mathematical physics, however, seems to have taken on a life of its own. The hot topic is the differential geometry of higher-dimensional spaces, with all the accompanying complications in describing how all these extra dimensions are supposed to exist without us seeing them in the three spatial dimensions of our reality. All this hasn't demonstrated its relevance for physics yet.

I don't complain that math is sometimes difficult. For instance, in continuum mechanics (a hot topic in the nineteenth century), you need nine numbers arranged in a matrix to describe how of a piece of material is deformed. The numbers contain information as to how the material can be rotated, stretched, or distorted in different directions. To get the pure rotation out of the deformation requires a difficult calculation called "polar decomposition." Nevertheless, we are talking about a problem that is clear to state, easy to visualize, and—physically—simple. In contrast, the modern math has become so convoluted that the physics often vanishes into a fog.

WHY WE DO NOT NEED FAIRY-TALE MATH

Since the mathematicians have invaded the theory of relativity, I do not understand it myself anymore.

—*Albert Einstein*

The three pillars of physics—electrodynamics, quantum mechanics, and the general theory of relativity—rest upon the differential equations named after Maxwell, Schrödinger, and Einstein. Einstein dabbled for years until he eventually discovered the ones needed for general relativity, while David Hilbert, a mathematician, deduced them right away in 1915. Hilbert outdid Einstein by using a tricky method developed by the mathematicians Euler and Lagrange more than 250 years ago.

Today's physicists, trained in sophisticated mathematics, try to imitate Hilbert's elegant way of working. But as an isolated technique, this is not physics yet.[1] Einstein's research consisted of linking the math with physical quantities, and thus producing testable predictions. This kind of work is not fashionable anymore, and this is why theoretical physics has so much trouble in predicting what can be observed.

If one day there should be a unified theory of physics, it must contain differential equations that embrace those of Einstein, Maxwell, and Schrödinger as approximations. However, the horror stories divulged by theoreticians about what kind of incredible new mathematical constructions ought to be invented for a unified theory are just a PR exercise intended to impress the public. It's the *physics* we don't yet understand.

And there isn't the slightest evidence that some higher-dimensional mathematical theory will be of any help.

CALCULATING UNTIL FIRST AID IS NEEDED

It makes my blood pressure rise when this type of gambling with physics is justified by referring to the founding fathers of modern physics such as Einstein. Just listen to the string enthusiast Brian Greene on YouTube, for example: "We believe we can realize Einstein's dream... Physicists are confident that we continue where Einstein has left off."

But to make Einstein the crown witness for complicated high-dimensional theories is utter insolence. He avoided any arbitrariness while searching for a unified theory in his lonesome, ascetic struggle with geometry. The modern theories, with their flamboyant extra dimensions, would have made him queasy. In contrast to these fantasies, Einstein kept his feet on the ground, looking for solutions to his equations that he hoped would correspond to two elementary particles, the proton and the electron.[2] I leave it up to you to imagine how he would have dealt with the particle blossom of the standard model and its extensions.

Today's theoretical physics is more and more a playground for mathematical consistency considerations. Theorists make arbitrary physical assumptions, and then rack their brains how to link fantasy A to fantasy B when there usually isn't an observational hint for either. It would seem that about 95 percent of theorists suffer from this disconnect from reality, which, interestingly, is about the same percentage as the dark substances in the universe—the former being obstinate, the latter opaque.

Instead, it would be more interesting to deal with unexplained observations that challenge the accepted theories. Many scientists simply do not feel the desire to bother with this kind of craftsmanship. Ask someone who is mathematically ironing out the infinities at black holes how he would measure the velocities in the Milky Way, or what the concentration of heavy elements in stars tells us about the age of globular clusters. There are plenty of baffling observations in the universe that await an explanation. But reality is really quite messy, so one does what one can, and that usually means calculating, no matter what.

Let's have a look at an excerpt from a ludicrous, though typical, paper of modern cosmology, "Inflation Dynamics and Reheating":

> We review the theory of inflation with single and multiple fields paying particular attention to the dynamics of adiabatic and entropy/isocurvature perturbations. [...] reheating and preheating after inflation providing a unified discussion of both the gravitational and nongravitational features of multi-field inflation...[3]

And so on. It means that the authors use a model with lots of arbitrary parameters to describe any observation, with calculations sufficiently inscrutable to be criticized. While discussing everything as vaguely as possible, they make sure not to hit any experimentally detectable limit. This is science today.

But you'll find hundreds of papers in this style on arXiv.org every month! Psychologically, this is what is called a collective displacement activity. And it is sociologically rewarded, since an error-free calculation can easily be published, however irrelevant it might be.

Fundamental science, though most often discovered by pure curiosity, sooner or later finds its use in technology. Every cell phone demonstrates Maxwell's groundbreaking discoveries. Such a real-world application would be unthinkable for the modern fantasies. The editor of *Medical Hypotheses*, in a critical paper dealing with general problems of science, commented: "Technology built using bogus theories will not work in the first place. And if it did happen to survive construction then would soon fall from the sky, collapse, or otherwise crash and burn."[4] Probably for good reason, "Inflation Dynamics and Reheating" takes place in an imaginary early universe.

HOMO SAPIENS MATHEMATICUS?

There is no belief, however foolish, that will not gather its faithful adherents who will defend it to the death.

—*Isaac Asimov*

Maybe in contrast to the views of the parties involved, abstract mathematics is not necessarily the field in which the intellect of *Homo sapiens* peaks. Fighting through arithmetical bushes and tracking the one-dimensional paths of logic didn't help the Neanderthals to survive. There is far too little evolutionary advantage for us to be talented in abstract math.

Rather, it is spatial perception, creating mental pictures of three-dimensional objects, and an extraordinary visual memory that distinguishes humans from other species. So, even if Nature were to be revealed in abstract math, we're holding a lousy hand of cards. Our strengths come into play where math connects with physical concepts, as for example when visualizing vector fields, wave functions, and curvature. That's where most of Nature's secrets were educed.

Einstein placed great trust in such intuition-based visual thinking, even though his calculations occasionally contained mistakes. He once mused about his work, "It's a simple story with the old one. Every year he revokes what he's written the year before."[5] In addition to having passionate curiosity, Einstein had great mental strength from the start. "God created the donkey, providing him with a thick skin." He did not care at all when a path he decided to follow was considered a dead end by everyone else. Without such stubbornness, Einstein would have never kicked off the scientific revolutions created by relativity or the light quanta.

> Most people say that it is the intellect which makes a great scientist. They are wrong: it is the character.
>
> —*Albert Einstein*

BEAUTY AND THE BRAIN

Einstein spoke several times about beauty in physical theories. The dangers of this argument are obvious. Recall that geocentric astronomy was believed to be true because its epicycles exhibited beautiful symmetry as circles. As Lee Smolin noticed, mathematical symmetry groups have slipped into an analogous role to epicycles today.[6] In contrast to the real symmetries of space-time, these constructs are completely abstract. Ultimately, this means that we are dismissing space-time and its geometry and building theories that go beyond them. But it is entirely possible that our problem is that we do not fully understand space and time yet. The modern approaches do not fail in an obvious way, because they still yield an approximate description of the observations, just as planets moving in circular paths did for the epicycle model. Today's mathematical constructs are impressed upon observations with the same brute force the poor circles were used back then. But Nature exacts revenge in both cases by necessitating more and more free parameters.

It is high time to bid goodbye to the ideological idolization of beauty. When Einstein and Dirac spoke of beauty, they had just climbed a high mountain with enormous effort, scientifically speaking, and could not find other words to describe the fantastic panorama their ideas revealed. But today "beauty" is invoked by mountaineering tourists equipped with thin ropes that propose the most expensive expeditions without even knowing which peak to climb. The pursuit of beauty has become a task of everybody's own taste, justifying even the greatest physical absurdities.

> I have little patience with scientists who take a board of wood, look for its thinnest part, and drill a great number of holes where drilling is easy.
>
> —*Albert Einstein*

THE GENIUSES ARE DEAD, LONG LIVE THE PRAGMATISTS

Do not follow where the path may lead. Go, instead, where there is no path and leave a trail.

—*Ralph Waldo Emerson*

Sadly, mainstream opinions also influence researchers who are actual critical thinkers in their own field. When meeting Julian Barbour in 2008, I was astonished at how he defended the "beautiful" symmetry concepts in elementary particle physics, and I was downright shocked that he, of all people, praised the "experimental evidence" for the standard model. This is the thinker who was ready to turn the standard model of cosmology upside down with his doubts about the Hubble expansion!

I objected in vain, using the arguments of Lee Smolin, who is horrified by the many free parameters of particle physics. On the one hand, Smolin is the "outlaw" in string theory, but on the other, he uncritically follows mainstream cosmology when he states that inflation has been "well-confirmed"—which would probably be shocking to Barbour.[7] Unfortunately, those who aren't experts in a specialized field usually do not doubt the majority's opinion, which then spreads unchallenged.

An enlightening account of the related problems of sociology in science is the book by Harry Collins, *Gravity's Shadow*. Collins particularly shows how group commitments, prestige, and personal opinions may determine what is established as fact.

SCIENCE NOW AND THEN: PEAK PERFORMANCE OR MASS SPORT?

One could talk about a collective intelligence of the big physics collaborations, which by far exceeds that of the individual.

—*Rolf Landua, CERN, Switzerland*

In order to be a distinguished member of a flock of sheep, one has to be, foremost, a sheep.

—*Albert Einstein, Bern, Switzerland*

We cannot underestimate how much the environment of science has changed. Individual thinking has been replaced by Big Science* with its dominating institutions, and that has had a great impact on theoretical physics. Although the founders of quantum mechanics communicated

---

* See the interesting book *Reflections on Big Science* by Alvin Weinberg.

a lot, the real leaps were made by Bohr, Heisenberg, de Broglie, and Schrödinger while working on their own. History shows no example of an important theory discovered by someone who conducted calculations while under the supervision of his professor. Einstein worked alone, and wrote:

"This kind of search in the dark with its tense desire, lasting for years, full of foreboding, with its exhausting change from aspiration to frustration and its final breakthrough to lucidity, all this you only know properly if you experienced it."[8] Einstein, however, in contrast to today's experts, was up-to-date in all the essential fields of theoretical physics.* Sometimes people say that physics just needs a new Einstein. But they are not thinking this through properly. "Fundamental" physics has become specialized in a manner that makes it impossible for an individual to stay on top of it all.

Instead, it is generally believed that modern research necessarily requires teamwork in the form of big collaborations. This might be the case for projects like the Large Hadron Collider or space telescopes. But with respect to theory, the teamwork argument would be more convincing if the collective efforts had actually yielded successes. Roger Penrose, in a chapter about fashion in his book *The Road to Reality*, thoughtfully comments:

In the present climate of fundamental research, it would appear to me much harder for individuals to make substantial progress than it had been in Einstein's day. Teamwork, massive computer calculations, the pursuing of fashionable ideas—these are the activities that we tend to see in current research. Can we see the needed fundamentally new perspectives coming out of such activities? This remains to be seen, but I am left somewhat doubtful about it.[9]

If you want to get an authentic impression of the work methods of theorists** you may compare Werner Heisenberg's autobiography *Physics and Beyond* with Alan Guth's *The Inflationary Universe*. Heisenberg, in great modesty, struggled with the fundamental questions of physics while hiking with his friends. In contrast, Guth's description of theorists who rival in tackling a fashionable problem, reminds me of lemmings who, having taken leave of their senses, collectively calculate themselves to death, but down the line remain convinced they are the brightest bulbs in the box.

---

* While dealing with the problem of expertocracy, the Spanish philosopher José Ortega y Gasset coined the beautiful term "The Barbarism of Specialisation."

** Also characteristic is Smolin's description of supersymmetry in chapter 5 of *The Trouble with Physics*.

BRAVE NEW WORLD—PHYSICS ROOTED OUT AND
REBUILT ANEW

Big Science, with its collaborations of thousands of physicists, dominates today's fundamental research. How did it develop? Is it still the same science? This is far from evident. At most times in history, a leading role in science and political power was closely related. After the heydays of physics in the early twentieth century, the barbaric Nazi regime also destroyed the heartland of science in Europe.

During a dinner in 1934 in Göttingen, Hitler's new minister for science asked the mathematician David Hilbert if his institute had suffered from the expulsion of the Jews. "Suffered?" retorted Hilbert. "No. It just doesn't exist anymore!" While many great minds of physics in Europe emigrated to the United States, they could not establish an American version of the prewar European scientific culture and its method of dealing with fundamental problems. A pragmatic approach began to dominate, and, of course, the importance of nuclear weapons contributed to this new orientation, as James Gleick, Feynman's biographer, pointedly spelled out: "[T]he scientists contemplating the state of theoretical physics descended into a distinct gloominess; in the aftermath of the bomb, their mood seemed postcoital."[10]

While the gain in military importance pushed physics toward Big Science, a dramatic methodological change in physics started in the middle of the twentieth century. Smolin describes:

> This style is pragmatic and hard-nosed and favors virtuosity in calculating over reflection on hard conceptual problems. This is profoundly different from the way that Albert Einstein, Niels Bohr, Werner Heisenberg, Erwin Schrödinger, and all the other early-twentieth-century revolutionaries did science. Their work arose from deep thought on the most basic questions surrounding space, time and matter, and they saw what they did as part of a broader philosophical tradition, in which they were at home.[11]

Postwar physics grew up without its old roots. Quantum electrodynamics had its successes, and it in turn became the blueprint for almost everything that followed under the name "quantum field theory." But in the last decades, its harvests have been rather poor. Obviously, the real problems cannot be solved by this pragmatic way of doing science. Not only does Smolin hit the nail on the head, but he is also impressively doubtful with respect to his own work—a virtue alarmingly absent among today's theorists:

> For the past twenty-five years we have certainly been very busy...But when it comes to extending our knowledge on the laws of Nature, we have made no real

headway... Unlike any previous generation, we have not achieved anything that we can be confident will outlive us.[12]

This is a surprising self-awareness in a researcher who has devoted a considerable amount of his life to that research! It seems, however, that the beginning of the crisis dates back about 60 years, if not to the unsolved riddles of quantum mechanics in the late 1920s.

With respect to the accomplishments of Einstein and his contemporaries, the physical theories in the new world of postwar science remain intellectual lightweights. It's like comparing the philosophical insights of Greek thinkers and Roman soldiers. Unfortunately, the parallel does not end there. Contemporary physics, with its epicycle-like standard model for particles and the scholasticism of string theory, seems to be regressing to the Dark Ages.

BACK TO THE ROOTS: THE THEORY IS ILL

It is difficult, if not impossible, for most people to think otherwise than in the fashion of their own period.

—*George Bernard Shaw*

Today's way of theorizing is a real matter of concern. The models are becoming ever more complicated and continue to produce new unexplained numbers, instead of diminishing the free parameters. The necessity of keeping in strict contact with observations is an essential element of science, but it is more and more dismissed these days. And calculating dominates over reflection.

As a result, arbitrary fashions such as supersymmetry or cosmic inflation dominate physics and live off naive extrapolations that reach absurdly many powers of ten. In the absence of observations, progress is usually established by acclamation, and wrong concepts can procreate over generations. Almost every theorist is convinced of the concept of mathematical symmetry constructs, the modern version of the medieval planetary circles. The insistence on the idea of an immovable Earth is worryingly similar to today's belief in the invariability of Nature's laws. Dirac was almost the only one reflecting upon this, and since his voice has died away, physics is governed by a pragmatic computing style, accompanied by huge amount of groupthink.

---

* That's the key message of David Lindley's excellent book *The End of Physics,* a razor-sharp analysis of how the nonsense gradually took over in modern physics.

Like a bad psychologist, we try to persuade Nature with our ever more complicated theories, instead of taking seriously the particular facts it is communicating to us. We have forgotten that understanding enigmatic observations is the primary task. Instead, we ask first if they conform to our theoretical junk, which we do not test anymore. In creating the myth of a unified theory,* physics disconnects from reality and produces speculation bubbles.

We have managed to transfer religious belief into gullibility for whatever can masquerade as science.

—*Nassim Taleb*

Chapter 20

BIG SCIENCE, BIG MONEY, BIG BUBBLES

WHAT'S WRONG WITH THE PHYSICS BUSINESS

Governments can delay an economic disaster by printing money. Physics, to avoid the bankrupting of its theories, can resort to experiments with ever-higher energies. But not every huge investment is healthy. Is big money essential for fundamental research? The answer seems obvious for Big Science experiments, which are expensive and often employ thousands of researchers. But does big money also help to answer big questions? Karl Popper, one of the most profound thinkers on science, points out:

"I think most physicists and biologists are very bright and conscientious. But they work under enormous pressure. This kind of pressure exists only after World War II, when science funding reached new dimensions."[1]

Today, funding has reached still greater dimensions. The large amounts of money being spent on research may, however, be cementing the stall of today's physics. Every project has to first get approval by a funding agency, and the agency's bureaucracies want to ensure that the money is not wasted. But it is precisely this requirement that practically stops groundbreaking new ideas by smothering creative and high-risk research. Those who usually vet funding applications are scientists who have achieved acclaim in their field, and they are likely to veto anything that threatens their particular sacred cow. Even branches of research based upon bogus theories may be kept going indefinitely by continuous transfusions of cash.

Truly original thoughts cannot develop where a monoculture of theoreticians, backed by the prestige of abundantly funded institutes, decides what is and is not interesting to study. I don't know how a perfect system can work, but right now the situation is worse than if we didn't pay theoreticians at all. I know that sounds provocative, but it really annoys me that so many young physicists are deterred by all the nonsense they're seeing published.

> Big science may destroy great science.
>
> —*Karl Popper*

THE DROPOUT PRIZE

An interesting idea for improving physics is to offer prizes for real and important discoveries. However, in a world where almost everything is headed toward privatization, this may not work. In the summer of 2012, Yuri Milner, a Russian billionaire, awarded his new "Fundamental Physics Prize" of US$3 million each to nine researchers who worked on string theory, cosmic inflation, and the holographic principle. Meeting the old-fashioned prerequisite of experimental verification demanded by the Nobel committee has eventually become obsolete.

What we are seeing in this new prize are the sumptuous omnipotence fantasies of a physics graduate school dropout.* It is tragicomic, really, that an intelligent entrepreneur such as Milner has, in effect, bought himself a future reputation as the foolish man who invested heavily in a bogus enterprise. However, Milner's decision to let the nine theorists who won the first awards select the future winners was a real masterpiece. Young researchers will now have to ask themselves whether they want to aspire to win the "Russian" or the true Nobel. This will eventually help to separate the sycophants of the string celebrities from those who want do real physics.

What might work instead is a modest subsistence for those who devote themselves to pursuing deep questions—a small salary for science monks.

---

*Not coincidentally, Milner threw another pile of prize money at LHC researchers the day after the Nobel ceremony. Hopefully, the Nobel Committee will not embarrass itself by doing the same in its future decisions.

INVESTING IN THE TALENT POOL OF PHYSICS

There are some options for the young physicist to make a lot of money fast. Physics journals often display large-size ads from well-connected head-hunting companies. With a master's degree in physics, you may get invited to a "getting to know you" sailing trip in the Bahamas, while the average graduate in economics is happy to merely sweat through a job interview. It is also not rare for physicists to be courted by the financial markets with lucrative contracts. This led the Nobelist Sheldon Glashow to seriously worry that theoretical physics may lose its new blood: "Promising graduate students had just left Harvard for Wall Street."[2] I believe however that one is *either* interested in how the market's laws make the world go round *or* in how Nature's laws make the world go round. It's not possible to have a stake in both.

Physicists working in the financial markets did not stand up against the collective idiocy that had taken place there, so I don't think they are such a great loss to physics. Just let them go! We might worry about more smart people who are prepared to gamble with money secured by taxpayers, but we should not hinder the migration of these talents to the investment banks.

FOOLED BY EXPERTISE

As soon as we abandon our own reason, and are content to rely upon authority, there is no end to our trouble.

—*Bertrand Russell*

The essential problem behind the financial crisis was the groupthink of the experts. Probably expert physicists do consider themselves to be immune to such a herd mentality. But, as a matter of fact, they have no chance to check the results of fields that are not their own. It is distressing how the specialists in physics are thoughtlessly parroting each other's opinions.* Questioning the model everyone is working with is utterly impossible once a new concept has been "established." Popper said: "The intellectuals make ideologies out of their theories. Unfortunately, also in physics many ideologies exist... Who does not follow the fashion easily drops out of the community members which are taken seriously."[3]

---

*It is another field, but just watch the film *12 Angry Men* to see how collective opinions may be prematurely shaped.

Often, the most meaningful progress in science was made by those who were considered cranks by the mainstream science community. A classic case of expert blindness occurred in geology in 1912. Alfred Wegener, a German meteorologist, drew a far-reaching conclusion from a simple observation. He pointed out that the Atlantic coastlines of South America and Africa looked like they should fit together. He was ridiculed by the famous geologists of the time because his idea didn't fit into their wrong theories. Only long after Wegener's death was the continental drift of tectonic plates accepted. What really fits this story is a quote of Jonathan Swift. "When a true genius appears, you can know him by this sign: that all the dunces are in a confederacy against him."

PUBLISH OR PERISH, BUT SCIENCE PERISHES BY PUBLISHING

None of the great discoveries was made by a "specialist" or a "researcher."

—*Martin H. Fischer, American author*

Although reviewers of scientific journals have provided well-documented examples supporting Jonathan Swift's observation, the scientific worth of a researcher is nowadays measured by the number of papers he or she has published. And more important than the paper's content is the "impact factor" of the journal, which computes the journal's reputation on the basis of the number of subsequent citations. In other words, being mentioned by other scientists has become the main yardstick by which the quality of a paper is measured. The self-fertilization of such counting is obvious. If everyone repeats a dumb idea about a problem in parrotlike fashion without solving it, it must mean that the idea is important.

Publications that are highly cited, such as the model of extra dimensions put forward by Lisa Randall and Alan Sundrum, are understandably called "influential papers"—they caused waves of a scientific influenza. But there are even worse epidemics. We can count more than 50,000 journal publications on supersymmetry. Experimental results are nonexistent, but in this herd, it is surely not difficult to find good-natured referees for the next round of funding decisions.

Sometimes I ask myself whether a fundamental theory that *works* would be able to sustain the thousands of physicists who currently play with their pet fantasies. The field of these experts would immediately lose its meaning, as soon as somebody made significant progress. Thus, string theorists and others should pay into an insurance fund, which could then cushion the unemployment effects once the promised Theory of Everything is discovered.

THE PRACTICE OF NONPEER NONREVIEW

I think there is such a thing as quality, but as soon as you try to define it, something goes haywire. You can't do it.

—*Robert Pirsig, American writer*

It is hard to say to someone face to face: Your paper is rubbish!

—*Max Planck*

The term "peer review" means that papers submitted to academic journals are reviewed anonymously by independent experts in the subject of the paper. Since the reviewers usually come from the same field as the author, every scientific fashion grown to a certain size develops its own dynamic, because critical questions beyond one's own nose become easier to ignore. Moreover, the adherents of the established models very efficiently seal themselves off from new ideas that threaten to shake their preconceptions. All this happens not necessarily in bad faith, because immersing oneself in somebody else's ideas is a rather unpleasant task.

For this reason, many referee reports are very superficial. Once I received almost overwhelming thanks from the editor of an astrophysical journal for a two-page statement I had written about one of their submitted manuscripts. I had indeed spent several hours reviewing it, but by no means did I check the validity of all the formulas.

Many reviewers content themselves with ascertaining whether the paper contradicts this or that experiment or perhaps even an established theory. That's if the decision has not already been made from checking the author's institutional background. A friend of mine who is working on gravitational physics in France, though his affiliation is in a different field, had a manuscript of his rejected several times. When he became a visiting scientist at a renowned university, his article was accepted at *Physical Review D* in a heartbeat.

THE RATING AGENCIES' RESPONSIBILITY—EVER HEARD OF THAT?

Richard Horton, the editor of the famous medical journal *The Lancet*, summarized his critique of the peer review system, which applies to physics as well:

We portray peer review to the public as a quasi-sacred process that helps to make science our most objective truth teller. But we know that the system of peer review is biased, unjust, unaccountable, incomplete, easily fixed, often insulting, usually ignorant, occasionally foolish, and frequently wrong.[4]

In 1996, the particle physicist Alan Sokal succeeded in playing a delight-ful hoax on the purveyors of the gibberish of postmodern philosophy. He submitted a completely nonsensical article entitled "Transgressing the Boundaries: Towards a Transformative Hermeneutics of Quantum Gravity" to the journal *Social Text*. The article was skillfully wrapped in technical jargon, and it was promptly accepted and published. The hoax, which Sokal announced immediately after the article's publication, generated consider-able mockery of the postmodernists by scientists. Maybe the editors of the physics journal *Classical and Quantum Gravity* laughed about this kind of thing as well, but in 2002 they surely did not. That's when they had to admit that an article they published by Igor and Grischka Bogdanov, two eccen-tric science popularizers, was blatant nonsense. It turned out later that the paper had been published in two other refereed journals, with almost the same preposterous content.

The only difference between this and the Sokal hoax was that the two discredited authors insisted that the paper was meant in all seriousness. (Embarrassingly, by the way, a Dirac medalist from MIT had particularly lauded the manuscript.)

QUALITY IN THE HAYSTACK—IN DUBIO PRO PUBLICO

The Bogdanov scandal was covered in the *New York Times* and in *Nature*, among other publications. However, the uproar over the Bogdanov paper should not be taken to mean that the accidental publication of baloney is somehow disastrous. There are probably thousands of published articles in peer-reviewed journals of a quality only slightly differing from the Bogdanov paper. Apart from some waste of time and paper, there wasn't much damage done.

It is, however, much more dangerous if an actual groundbreaking idea is considered to be absurd and is then rejected by the editors. Long ago, *Nature* turned down an article that later led to the Nobel Prize for Medicine in 1953, and there are plenty of examples of this kind of mistake in the his-tory of scientific publication.[5] This is similar to how in criminal law, the wrongful conviction of a scientific visionary is a lot worse than a handful of "crooks" getting away unpunished just because they got published when they shouldn't have. Maybe science should endorse the judicial policy *in dubio pro reo*—when in doubt, decide for the accused.

By the way, many believe that this has already been achieved through online publication on the e-print server ArXiv.org. According to the web-site, it provides the opportunity to publish at "low costs" (free for submit-ters), but there also happens to be a corresponding low level of fairness and transparency. A subtle, but nonetheless rigid, censorship called *moderation* has been installed that, as ArXiv proudly states, works "effectively behind the scenes."

The effectiveness works like this. Once you write critically about string theory, for instance, your name goes on a blacklist, which means your papers are either blocked or shifted to the low-rank "general physics" category. Curiously, the most prominent victim of this blacklisting is the 1973 Nobel laureate in Physics, Brian Josephson. No matter how far he might have drifted toward esoteric topics, does one really have to protect ArXiv readers from his articles? The argument that this screening of articles is needed because of limited resources is just ridiculous in an era of terabyte hard drives. The only reasonable way to avoid the suppression of ideas is to completely disentangle publication from evaluation.* At the very least, unknown and nonmainstream scientists could be given a quota of the annual number of articles they can publish rather than being blocked. But this appears to be an unthinkable solution for the ArXiv gatekeepers.**

> It becomes more indispensable than ever to preserve the freedom of scientific research and the freedom of initiative for the original investigators, because these freedoms have always been and will always remain the most fertile sources for the grand progress of science.
>
> —*Louis Victor de Broglie, Nobel laureate, 1929*

NOBODY BELIEVES THE THEORY, EXCEPT FOR THE THEORIST

It is well known that "risky" ideas that run contrary to established views are hard to publish, while boring, technical papers are usually waved through by the reviewers. As long as the paper doesn't have mistakes, it doesn't necessarily have to contain an idea.

Open peer review, instead of the anonymous system, is one idea, and another one is to trade scientific theories analogous to the financial markets—whoever considers something to be hokum that later turns out to be a big discovery will have to pay for that wrong assessment. It's a neat proposal, but if the high-trading successes of string theory are ever stored in a "bad bank" like securitized mortgage loans, should we then have to listen to those involved claiming that no one saw the bubble coming, as was the case in the American congressional hearings? A well-meant proposal to allocate a certain percentage of funds for unconventional ideas has also

---

* This is proposed, among some other reasonable ideas, in an article entitled "Scientific Utopia" (arXiv:1205.4251).

** In reaction to this, the completely open site vixra.org has been created.

been made, but it will presumably end like the industrialized countries' promise to supply developmental aid. Several institutions, such as FQ(Xi), have been founded that claim to support "risky," speculative research—and were promptly infiltrated by mainstream guardians of the status quo. Whose job is it to distinguish nonsensical speculation without observation from interesting ideas about unexplained observations? That's where asylum is really needed.

EVERYBODY BELIEVES THE EXPERIMENT, EXCEPT FOR THE EXPERIMENTER

> The processes through which...experimental procedures are evaluated in the light of group commitments...surely deserve the closest attention.
>
> —Andrew Pickering, science historian

An experiment is usually considered to be an impartial judge that determines whether a theory is correct or not. The history and sociology of science have long since shown that such an unquestioning belief in experimental results—called "naïve realism"—is plain wrong. A particularly enlightening discussion of this issue is given in the book *Constructing Quarks* by Andrew Pickering, a former particle physicist. Pickering showed that there is an intimate relationship between experimental practice and theoretical belief that has often been overlooked. Scientists tend to prefer data that conforms with the accepted theories and adjust their experimental techniques to "tune in" on phenomena consistent with expectations.

Every practitioner of science knows, however, that the pressure to publish often leads to squeezing as much out of the data as possible. It is easy to fool ourselves with too-simple presumptions that lead to the desired outcome. But since only a few experts are qualified to make objections*—a rare case—the results attract attention. The authors eventually die, but the paper lives on. No one likes to rain on a colleague's parade, and there are certainly no laurels to be gained by it.

A friend of mine once took the time to pore over the data evaluation of a satellite flight, for which an Italian physicist had claimed an utterly implausible accuracy. He submitted a commenting paper to the journal that had published the satellite article.[6] Although the reviewers admitted that my friend was correct, his commenting paper was only accepted after long discussion. One referee even countered that the errors in the criticized

---

* It is certainly worthwhile to wonder who would be qualified to criticize a paper by the big particle physics collaborations with several thousands of authors.

paper were evident and didn't need to be pointed out! Editors dislike such an embarrassing step backward.

Ironically, scientific flagships such as *Nature* and *Science* are not immune to the publication of papers of dubious merit. Some articles are simply not detailed enough for an objective review. This means that if you interpret scratches on a rock on Mars as traces of aliens, you might well be accepted for publication. If, however, your article is about the questionable analysis that fails to support the alien conclusion, it is most likely to be rejected as not being of sufficient *general interest*. There is a motivation for career-oriented scientists to focus their efforts on packaging the content.

KNOWLEDGE BANKS AND THEIR CREDIT RATING

It is especially dangerous when the raw data from extensive observations is only available to a single research group, which carries out the entire evaluation. Imagine if some big experiments on which physics has been relying for decades contain unknown systematic errors. That is a nightmare. Just as in the case of Einstein and de Haas mentioned in chapter 12, where their desire to comply with the expectations made them blind to errors, this doesn't mean anyone had bad intentions.

Remember that in 2012, researchers of the OPERA collaboration, after an elaborate analysis, claimed to have measured neutrinos traveling faster than the speed of light. It turned out to be a problem with a defective electrical plug, instead. Scientists involved in the big detector experiments at colliders will claim that they work more cautiously, but the complexity, particularly where the computer simulations are removing the background noise, has become rampant.

That being said, it is obviously not realistic to take publication in a scientific journal as the sole evidence for the correctness of a result. In 2008, at a lecture in Munich, the Nobelist Murray Gell-Mann said, "I once published a paper together with Feynman which contradicted eight published experiments. They were all wrong." This was merely a joke, similar to a witticism in cosmology: "No theory should reproduce all data, since some is surely wrong." But modern observations have become far too complex to be checked independently by a single person. Publication in a journal is like being ranked by a credit rating agency, and a paper in *Physical Review* corresponds to an AAA rating. The researcher's publication "win" is like a bonus for fund managers, which does not take long-term risk into account. Once science starts to build upon dubious calculations, a collapse is inevitable. Piling up badly tested results is a lot like piling up questionable derivatives on balance sheets. At this point, physicists are still buying everything off each other. But remember how stock analysts and financial experts demonstrated their blindness—in some cases, willful blindness—right before the 2008 crash. Only the future will tell whether we are sitting on a huge heap of toxic papers or not.

> What is the chance that a person who notices an important discrepancy in a scientific announcement has the opportunity to check it out at the level of the primary data?
>
> —*Halton Arp*

BETWEEN SKYLLA AND CHARYBDIS: CONSPIRACY THEORIES DO NOT HELP

Am I too skeptical? Long ago, I had the opportunity to briefly talk to Bobby Fischer, the world chess champion in the early 1970s. The unfortunate genius had convinced himself that the world is directed by a globally operating Zionist conspiracy that controls all governments and media. When I carefully remarked that I decided to study physics for the very reason that it deals with objectively verifiable facts, he suddenly came up with warnings, advising me to study some "enlightening" books, which I have not read to this day.

If you doubt *everything* you haven't seen with your own eyes, you are easily trapped in conspiracy theories. Popper's criterion of falsifiability applies here—the assumption that all information surrounding us has been manipulated can, unfortunately, not be disproved. So how do you find the right amount of skepticism? If you find my views about the current state of affairs in physics to be too mistrusting, please read *A Different Universe* by the Nobelist Robert B. Laughlin. He is very concerned about the weaknesses undermining science right now. In contrast, the majority of physicists will have little doubt about "established" results. It has, however, become a system based on mutual confidence. This is also a lot like the financial markets, and we know what happens when confidence fails.

MODERN REPRODUCIBILITY—CLICKING THROUGH RAW DATA

> Do not hope without doubting and do not doubt without hoping.
>
> —*Seneca, ancient philosopher*

To believe or not to believe...that is not the way to resolve a scientific conflict. What we need is a true transparency of observational data. Empirical science also means that somebody must be able to reproduce the results, and that cannot happen if only one group runs the big expensive experiments, analyzes the results themselves, and keeps the raw data as a form of proprietary knowledge.

The experiments you read about in early twentieth-century journals are often the kind that would not be difficult to conduct in your garage. Since then, the size and expense of the experiments has exploded, but the old-fashioned descriptions on a piece of paper are still there. Just try to understand the setup of an experiment from a journal's description. Photo and video documentation could partly help, but the real problem lies in the computer programs that analyze the more complex experiments, not to mention the Big Science giantism. Since the code is usually inaccessible to the public, it is practically impossible to reproduce the results. One of the basic requirements of scientific method is missing here.

FIGHTING FRAGILITY

Taking a very general perspective, science has to change its methodology. What is needed is a new standard for evaluating data. Put the raw form on the Internet, so that each step can be scrutinized independently. Everything would, of course, be documented in a publicly accessible source code. Public access should become the norm, particularly since most of the funding comes from public coffers.

If different researchers work on raw data reduction and on the final interpretation, as is the case with the Sloan Digital Sky Survey (SDSS), this can only have a positive effect on an agenda-free analysis. Critical observers should be able to try out their own modifications of the program. This would be real reproducibility. And who knows? We might even find some undiscovered treasure, as various researchers take a fresh look at the data and analyze it under new aspects. In astronomy, such a new way of science seems to have a fair chance of succeeding, and we may avoid the road toward new epicycles that the standard model seems to be favoring.

In contrast, particle physics has no public data that anybody, except high-energy physics experts, can make sense of. The field seems to have piled up too many poorly understood facts. It is an increasingly unstable structure held together mostly by the belief of its adherents. But the continuous inflation of arbitrary concepts has long undermined its credibility. The longer this situation will last, the harder the crash will be. One day it will be suddenly clear to the majority that this is not Nature's picture but human tapestry.

To summarize, besides the obviously absurd theories such as strings, supersymmetry, and cosmic inflation, there is a deep underlying crisis of allegedly "real" physics that is not widely perceived to date. Overly complicated models seem to be supported by evidence that has gradually shifted from genuine observation to a social consensus. Physics has become a fragile building that sooner or later will collapse.

Chapter 21

OUTLOOK

GET PREPARED FOR THE CRASH

I had a lot of fun writing this book, but I realize that my portrayal of physics is incomplete. The fascinations of the universe are really multifaceted, and I'd like to encourage you to explore them in depth for yourself. I also could not cover many absurdities theoretical physics has brought forward in the past decades and only illustrated a few particularly egregious ones. Hopefully, no one feels overlooked. Yet the critical comments of my rather sympathetic proofreaders assure me that experts, unhappy with having their ox gored, will have plenty of targets for picking apart my critique of present-day physics. I'm really curious!

You may have found a bit of my skepticism to be exaggerated. I would, by the way, be surprised myself if I wasn't mistaken in any of my assessments. I make no claim to perfection, but one thing is certain. The highly specialized pursuits of fundamental physics are seriously in danger of becoming dependent on belief.

But, you may ask, isn't it presumptuous to judge or to put in doubt theories without being a specialized expert oneself? Think about it. As a scientist, I find it sobering to see how modest our knowledge of Nature still is; at the same time, science provides us with tools to simplify life once we reflect a little.

A physicist does not have to argue about perpetual motion in detail, because due to the conservation of energy, it won't work in any case. Nor will he or she believe that you can become rich in a casino with an ingenious method of placing your chips, however sophisticated, beautiful or complex the system may be. And likewise, science cannot work in the long run without experiments. Whatever debauchery of seductive mathematics string theory will come up with, you don't have to study theology in order to understand that it isn't science.

The mortgage specialist at your bank can regale you with numbers for hours, but if your question about the annual percentage rate isn't answered convincingly, forget about signing a contract. The question to be posed to the advocates of the standard model of elementary particles is, "How many free parameters do you have?" Well, forget about signing on with that theory, then.

Astronomy distinguishes itself from astrology by making astrology's absurdities apparent. The zodiac signs have long since shifted, the influence of a midwife is stronger than that of Uranus, and biology has some tiny objections as well. Analogously, we must challenge cosmic inflation, pointing to its extrapolations from the very moment of the Big Bang. Feel free to call the absurd "absurd," even if there are a great many people earnestly calculating fluctuations or ascendants.

If you see fraud and don't shout fraud, you are a fraud.

—*Nassim Taleb*

THE DAY AFTER

Science means, after all, not being a sucker. It is healthy skepticism that upholds physics. You can doubt whether quantum gravity may ever predict a number like 10^{40}, just as a meteorologist doubts weather lore. And the weather prognostications of the *Farmers' Almanac* are a reasonable and modest thing compared to the grand extrapolations of particle physics. We would mistrust a doctor who promises to come up with a universal cure for everything, but nevertheless many believe the Theory of Everything to be on its way. Physics has become more careless than any other science, so please, do doubt!

Unfortunately, one has to doubt as well the allegedly well-tested experimental results supporting the standard models. More than ever, it is necessary to go back to the raw data, making every step transparent and asking, "Is this really the best, simplest, and only possible explanation?"

The standard models of cosmology and particle physics are ailing, but theoretical physics overall is in a much worse shape. There has been no real output at all in the past decades. Strings, inflation, and multiverses have formed gigantic bubbles of speculation that can barely hide their ridiculously empty content. It is likely that theorists will aggressively defend their bogus fantasies for some time yet. But the higher-dimensional rigmarole will eventually be identified by the public and deflated. When the fantasies end in smoke, physics will be left without any credible theory. It will have lost its credibility.

In economics, this is the point where the crisis hits the real economy. Physics will be confronted with demands to know what we got out of all that time and money spent on Big Science experiments over the past half a century. The overly complicated constructions of the standard models will then get no credit either. This is a bankruptcy.

And this is how science has always worked. The crisis, after all, is also an opportunity. We physicists have a great responsibility to prevent our science from being liquidated. Leaving the world to ideology, religion, and antiscience would be the real catastrophe. Therefore, we must continue to use our minds. It is *real* physics what we need.

APPENDIX

THE DIRAC LARGE NUMBERS HYPOTHESIS

Before starting, you have to be aware that due to the huge numbers involved in the calculations, factors such as 10 or even 100 do not really matter if we are only considering orders of magnitude. However this coincidence is relatively accurate:

$$c\, m_p\, r_p \approx h.$$

multiplying both sides with c

$$c^2 m_p r_p \approx hc \approx 137 \frac{e^2}{2\varepsilon_0},$$

when using the inverse of the fine structure constant $\frac{e^2}{2hc\varepsilon_0} = 137$. Then dividing by G and m_p^2, followed by expanding the r.h.s. fraction by 2π, it follows

$$\frac{c^2 r_p}{G m_p} \approx 860 \frac{e^2}{4\pi\varepsilon_0 G m_p^2}$$

Besides the number of 860 on the r.h.s. we precisely find the ratio of the electrical to the gravitational force of two protons, about $1{,}2 \times 10^{36}$, while the l.h.s. can be transformed using Eddington's observation $\frac{c^2}{G} \approx \frac{M_u}{R_u}$:

$$\frac{M_u}{m_p} \frac{r_p}{R_u} \approx 860 \times 1{,}2 \times 10^{36} \approx 10^{39}$$

Since $\frac{R_u}{r_p} \approx 10^{41}$, Dirac's observation regarding the number of protons in the universe $\frac{M_u}{m_p} \approx 10^{80}$ follows. Thus Dirac's large numbers are closely related to $cm_p r_p \approx h$.

THANKS

The biggest thanks go to my coauthor, Sheilla Jones, for her patient listening, her very clear advice, and for the heroic struggle to transform my awkwardly phrased sentences into a pleasing read.

I am grateful that my wife and children continued to have patience with me during this third book project, and I would also like to mention my parents, to whom I owe a lot. As far as physics is concerned, the greatest source of support and encouragement over the years has been Karl Fabian.

This project wouldn't have been realized without the commitment of our agent, Ethan Ellenberg. Our special thanks go to our editor, Luba Ostashevsky, Laura Lancaster, and their team at Palgrave Macmillan for their patience and advice. We further thank Anselm Coogan and Nicholos Wethington for contributing to the translation, and Kris Krogh, Natalia Kiriuscheva, Wolfgang Kundt, Pavel Kroupa, and Peter Thirolf for their comments on the manuscript.

PERMISSIONS

Fig.		Permission granted by
1	Coma cluster	NASA
1	Fritz Zwicky	Zwicky Foundation, Glarus (CH)
2	Radio telescope Wettzell	Authors' photograph
3	Multiwavelength Milky Way	NASA
4	Letter Allan Sandage	Allan Sandage, Rudy Schild
5	Globular cluster Namibia	Stefan Geier
6	Einstein's apartment	Authors' photograph
7	Rotation curve	Van Albada, Sancisi, Petrou and Tayler, Phil. Trans. Roy. Soc. 320 (1986), 1556, Royal Society
7	Galaxy NGC 3198	John Vickery, Jim Matthes/Adam Block/NOAO/AURA/NSF
8	Galaxy distribution	The 2dF Galaxy redshift survey team, www2.aao.gov.au/2dFGRS/
9	Cosmic Microwave Background	NASA+WMAP Science team
10	Evolution of the universe	NASA+WMAP Science team

NOTES

CHAPTER 1

1. J. D. Anderson et al., *Phys. Rev.* D65 (2002): 082004, arXiv:gr-qc/0104064.
2. A. Aguirre et al., *Class. Quant. Grav.*18 (2001): R223-R232, arXiv:hep-ph/0105083.
3. J. Binney and S. Tremaine, S. *Galactic Dynamics* (Princeton, NJ: Princeton University Press, 2008), 635.
4. Smolin, 13.
5. Wilczek et al., *Phys. Rev.* D 73 (2006): 023505, arXiv:astro-ph/0511774.
6. Lewis Ryder, *Quantum Field Theory* (Cambridge: Cambridge University Press, 1996), 3.
7. Greene, 19.
8. Günther Hasinger, *Das Schicksal des Universums* (Munich: Beck, 2007), 60.
9. C. P. Burgess and F. Quevedo, *Scientific American,* November 2007, 53ff .

CHAPTER 2

1. HORIZONS System, http://ssd.jpl.nasa.gov/?horizons.
2. See International Earth Rotation and Reference Systems Service website, www.iers.org and Helmholtz Centre Potsdam GFZ German Research Centre for Geosciences, www.gfz-potsdam.de.

CHAPTER 3

1. Planck collaboration, arXiv.org/abs/1303.25062.
2. See Harvard-Smithsonian Center for Astrophysics, www.cfa.harvard.edu/~huchra/ and "The Cepheid Distance Scale: A History," at www.institute-of-brilliant-failures.com/.
3. Singh, 56.
4. C. Conselice, *Scientific American*, February 2007, 34.

CHAPTER 4

1. Einstein, 136.
2. J. D. Barrow, *American Journal of Physics* 57, no. 6 (1989).
3. Robert Debever, *Letters on Absolute Parallelism* (Princeton, NJ: Princeton University Press, 1979).
4. Feynman (Lectures II), chap. 28.
5. Chen et al., *Nature* 396 (1998): 653–655, arXiv:physics/9810036.
6. LIGO collaboration, arXiv.org/abs/0909.3583.
7. B. S. Sathyaprakash and B. F. Schutz, *Living Reviews in Relativity* 12 (2009), livingreviews.org/Articles/lrr-2009–2.

CHAPTER 5

1. W. Michaelis et al., *Metrologia* 32 (1995): 267–276.
2. See, for example, A. Unzicker, arXiv.org/abs/0702009, 3.

3. Gundlach et al., *Phys. Rev. Lett.* 85 (2000): 2869–2872, gr-qc/0006043, Quinn et al. *Phys. Rev. D* 87 (2001): 111101–1.
4. H.V. Parks et al., *Phys. Rev. Lett.* 105 (2010): 110801, arXiv:1008.3203; L. Jun et al., *Phys. Rev. Lett.* 102 (2009): 240801–240801.4 , P. J. Mohr et al., *Rev. Mod. Phys.* 84 (2012): 1527–1605, arXiv:1203.5425.
5. D. R. Mikkelsen et. al., *Phys. Rev. D* 16 (1977): 919.
6. F. Palmonari et al., *Phys. Rev. D* 64 (2001):082001.
7. J. Thomas et. al., *Phys. Rev. D* 63 (1989):1902.
8. M. Ander et.al., *Phys. Rev. D* 62 (1989): 985.
9. R. Genzel et al., *Rev. Mod. Phys.* 82 (2010): 3121–3195, arXiv:1006.0064.
10. See www.physik.uni-augsburg.de/annalen/history/einstein-papers/1916_49_769–822.pdf, 771.
11. D. Sciama, *Mon. Not. Roy. Astron. Soc.* 113 (1953): 34.
12. J. Barbour, *Class. Quant. Grav.* 20 (2003): 1543–1570, gr-qc/0211021.
13. See "Feynman: Inertia and Fathers," YouTube, http://www.youtube.com/watch?v=HgAQV 05fPEk.
14. M. Mamone Capria, *Physics before and after Einstein* (IOS Press, 2005), 156.
15. J.-P. Uzan, *Rev. Mod. Phys.*75 (2003):403, arXiv:hep-ph/0205340.
16. D. J. Stephenson, *Astron. u. Geoph.* 44 (2003): 2. 22.
17. F. S. Accetta et al., *Phys. Lett.* B 248 (1990): 146.

CHAPTER 6

1. A. Bosma, arXiv.org/abs/astro-ph/0312154.
2. M. Persic and P. Salucci, arXiv.org/abs/astro-ph/9502091.
3. G. Gentile et al., *Mon.Not.Roy.Astron.Soc.*375 (2007):199–212, arXiv:astro-ph/0611355.
4. Sanders, 68.
5. R. Scarpa et al., arXiv.org/abs/0707.2459.
6. P. Kroupa, *Pub. Astron. Soc. Austr.* 29(4) (2012): 395–433, arXiv.org/abs/1204.2546. (Falsification of Dual Galaxy Dwarf Theorem.)
7. A. Aguirre et al., *Class. Quant. Grav.*18 (2001): R223-R232, arXiv:hep-ph/0105083.
8. Sanders, 165.
9. U. Sawangwit and T. Shanks, *Astron. Geophys.* 51 (2010): 5.14 – 5.16.
10. J. Hogan, *Nature* 448 (2007): 240–245.
11. A. Aguirre et al., *Class.Quant.Grav.*18 (2001) :R223-R232, arXiv:hep-ph/0105083.
12. R. H. Sanders and S. S. McGaugh, *Ann. Rev. Astron. Astrophys.* 40 (2002): 263–317, arXiv:astroph/0204521.
13. G. Gentile, arXiv.org/abs/0805.1731.
14. K. Krogh, arXiv.org/abs/astro-ph/0409615, A. Levy et al., *Adv. Sp. Res.* 43 (2009): 1538–1544, arXiv:0809.2682.
15. R. Scarpa, arXiv.org/abs/astro-ph/0601581.
16. Smolin, chap. 13.

CHAPTER 7

1. G. D. Bothun, *Scientific American*, February 1997, 56–61.
2. J. Sellwood et al., arXiv.org/abs/astro-ph/0009074.
3. W. Evans, arXiv.org/abs/astro-ph/0102082.
4. M. Disney et. al., *Nature* 455 (2008):1082–1084, arXiv:0811.1554.
5. M. Disney, *Gen. Rel. Grav.* 32 (2000):1125–1134, arXiv:astro-ph/0009020.
6. G. Besla et al., *Astrophys. J.* 668 (2007):949–967, arXiv:astro-ph/0703196.
7. M. Metz et al., *Astrophys. J.* 697 (2009): 269–274, arXiv:0903.0375; see also M. Pawlowski et al., *Mon. Not. Roy. Astron. Soc.* 424 (2012), 80–92.
8. Parker, chap. 10, 222.
9. A. Tasitsiomi, *Int. J. Mod. Phys.* D12 (2003): 1157, arXiv:astro-ph/0205464.
10. "DarkMatter vs. Dark Energy–Leonard Susskind," YouTube, www.youtube.com/watch?v=3SiGujnfDVc.
11. G. de Vaucouleurs, *Science* 167 (1970): 1203ff .
12. V. De Lapparent, M. Geller, and J. Huchra. "A Slice of the Universe," *Astrophys. J.* 302 (1986): L1–L5.

13. Cornell, chap. 3 (M. Geller), 72.
14. I.D. Karachentsev et al., *Astron.Astrophys.* 389 (2002): 812–824, arXiv:astro-ph/0204507.
15. P. J. E. Peebles, *Nuovo Cim.*B122 (2007):1035–1042, arXiv:0712.2757.
16. F. Sylos Labini, *Class. Quant. Grav.* 28 (2011): 164003, arXiv.org:1103.5974 and *Europhys. Lett.* 96 (2011): 59001, arXiv: 1110.4041.
17. M. Disney, *American Scientist* 95, no. 5 (September-October 2007), www.americanscientist.org/issues/pub/modern-cosmology-science-or-folktale.
18. Disney, *Nature* 263 (1976), 573.
19. N. Nabokov et al., *Astrophys.Bull.*63 (2008):244–258, arXiv:0901.0405; M. López Corredoira, arXiv.org/abs/1002.0525.
20. G. de Vaucouleurs, *Science* 167 (1970): 1203ff .
21. *New Scientist*, May 2004, cosmologystatement.org.
22. M. López Corredoira, *Int. J. Astron. Astrophys.* 1 (2011):73–82, arXiv:0910.4297.
23. M. Baldi, *Mon. Not. Roy. Astron. Soc.* 412 (2011):L1, arXiv:1006.3761.

CHAPTER 8

1. Y. Baryshev, arXiv.org/abs/0810.0153.
2. "PowerSpectrumExt," *Wikipedia, The Free Encyclopedia,* en.wikipedia.org/wiki/File:PowerSpectrumExt.svg.
3. D. J. Eisenstein, "Dark Energy and Cosmic Sound," www.cfa.harvard.edu/~deisenst/acoustic-peak/acoustic.pdf
4. F. Sylos Labini et al., *Astron. Astrophys.* 505 (2009): 981ff, arXiv:0903.0950.
5. Interestingly, the significance of the effect seems to have decreased. See E. A. Kazin et al., *Astroph. J.* 710 (2010): 1444–1461, arXiv:0908.2598.
6. H. Liu et al., *Mon. Not. Roy. Astron. Soc. Lett.* 413 (2009): L96-L100, arXiv:1009.2701.
7. D. Spergel et al., *Astrophys.J.Suppl.148* (2003):175–194, arXiv:astro-ph/0302209.
8. P.-M.L. Robitaille, *Prog. in Phys.* 1 (2007): 3–18, www.ptep-online.com/index_fi les/2007/PP-08–01.PDF.
9. H. Liu and T.-P. Li, arXiv.org/abs/1203.5720.
10. S. S. McGaugh, arXiv.org/abs/0707.3795.
11. Paul Murdin, *End in Fire* (Cambridge: Cambridge University Press, 1990), chap.9.
12. M. Disney, *Gen. Rel. Grav.* 32 (2000):1125–1134, arXiv:astro-ph/0009020.
13. A. Loeb, *Scientific American*, November 2006, 47.
14. Collins, 510.

CHAPTER 9

1. V. Springel et.al., *Nature* 435 (2005): 629–636.
2. Sanders, 118.
3. F. Wilczek, arXiv.org/abs/0708.4361.
4. M. Disney, *Gen. Rel. Grav.* 32 (2000):1125–1134, arXiv:astro-ph/0009020.
5. A. Belikov and D. Hooper, *Phys. Rev.* D80 (2009):035007, arXiv:0904.1210.
6. M. López Corredoira, arXiv.org/abs/0812.0537.
7. Magueijo, 137.
8. Moustakas et al., arXiv.org/abs/0902.3219.
9. Horgan, chap. 8, 226.

CHAPTER 10

1. Cornell, chap. 5, 105.
2. Cornell, chap. 5, 105.
3. Guth, in Cornell, chap. 5, 110.
4. "Parallel Universes," BBC, http://www.bbc.co.uk/science/horizon/2001/parallelunitrans.shtml.
5. Penrose, 754.
6. Ibid.,756.
7. Ibid., 753.
8. BBC Online, news.bbc.co.uk/2/hi/science/nature/1270726.stm.
9. Magueijo, 125.

10. R. Dicke, *Rev. Mod.Phys.* 29 (1957): 363.
11. www.physik.uni-augsburg.de/annalen/history/einstein-papers/1911_35_898–908.pdf
12. The Canadian physicist John Moffat had anticipated Magueijo's theory in 1992, but got proper recognition for it much later.
13. G. F. R. Ellis, *Gen.Rel.Grav.* 39 (2007): 511–520, arXiv:astro-ph/0703751.
14. J Barrow et al., *Gen.Rel.Grav.* 29 (1997): 1503–1510, arXiv:gr-qc/9705048.
15. P. Steinhardt, *Scientific American*, April 2011, 36–43.
16. Cornell, chap. 5 (A. Guth), 122.

CHAPTER 11

1. Davies and Brown, 183.
2. A. Staruszkiewicz, *Concepts of Physics* 1 (2004):169, merlin.fi c.uni.lodz.pl/concepts/2004 _1_2/2004_1_2_169.pdf.
3. A. Ashtekar, *Gen. Rel. Grav.* 41 (2009): 707–741, *Gen. Rel. Grav.* 41 (2009): 707–741, arXiv:0812.0177.
4. J. Erdmenger et al., *Phys. Rev. Lett.* 98 (2007): 261301, arXiv:0705.1586.
5. J. Casares, arXiv.org/abs/astro-ph/0612312.
6. Interview on *Fox News*, YouTube, www.youtube.com/watch?v=V-vG7o9Ioo4.
7. Segrè, 293.

CHAPTER 12

1. Howard Georgi, *Lie algebras in Particle Physics* (Reading, Mass.: Benjamin/Cummings, 1982); arXiv.org/abs/0805.3500, 9.
2. Feynman (*QED*) chap. 4, 142.
3. L. Wang et.al., *Phys. Rev.* D78 (2008): 013003, arXiv:0804.1779.
4. See the interesting comment by Schroer, arXiv.org/hep-th/9410085.
5. Feynman (*QED*), chap. 4, 149.
6. Gleick, 282.
7. Penrose, chap. 25.7.

CHAPTER 13

1. Segrè, 291.
2. See Feynman (*Lectures*) chap. 28; Landau, § 75.
3. See Pickering, diagram, 259.
4. R. Pohl et al. *Nature* 46 (2010): 213–217.
5. Feynman (QED), chap. 4, 148.
6. Dürr et al., *Science* 322 (2008): 1224–1227
7. Lindley (EoP), 113.

CHAPTER 14

1. *Not Even Wrong* (blog), http://www.math.columbia.edu/~woit/wordpress/.
2. "The Smell of SUSY," *Not Even Wrong* (blog), May 16, 2012, http://www.math.columbia.edu/~woit/wordpress/?p=4696.
3. Feynman (QED*)*, 150.
4. Feynman (QED), 129.
5. P. A. M. Dirac, *Proc. R. Soc. Lond.* A 165 (1938): 199 208.
6. Pascual Jordan, *Schwerkraft und Weltall* (Braunschweig: Vieweg, 1955), chap. 36, 247.
7. It's a pleasure to listen to Dirac's talk on the subject: see www.paricenter.com/library/download/dirac01.mp3.
8. J.-P. Uzan, *Rev.Mod.Phys.*75 (2003): 403, arXiv:hep-ph/0205340.
9. Lisa Randall, "Theories of the Brane," www.edge.org/3rd_culture/randall03/randall03_print.html.
10. "Hiding in the Mirror," *Not Even Wrong* (blog), October 18, 2005, www.math.columbia.edu/~woit/wordpress/?p=281.
11. Kapner et al., *Phys. Rev. Lett.* 98 (2007); 021101.
12. Smolin, xvi.
13. Lisa Randall, *Knocking on Heaven's Door* (New York: HarperCollins, 2012), 266ff.
14. Lawrence Krauss, *A Universe out of Nothing* (New York: Free Press, 2012),73f.

15. Ibid., 128f.
16. Johann Von Grolle and Rafaela von Bredow, "Die Welt ist bizarre," *Der Spiegel*, March 14, 2005, www.spiegel.de/spiegel/print/d-39694676.html.
17. Translated from *Der Spiegel* 1(2008), 31.12.2007, 120.
18. Horgan, 101f.

CHAPTER 15

1. Greene,11. See also Brian Greene, "The Elegant Universe Superstrings, Hidden Dimensions, and the Quest for the Ultimate Theory: An Excerpt," www.byronevents.net/science/elegant.htm.
2. G. Taubes, *Science* 23 (July 1999), www.sciencemag.org/content/285/5427/512.summary.
3. H. D. Zeh, www.rzuser.uni-heidelberg.de/~as3/M-Th eorie.html.
4. Smolin, xv.
5. "Number Theory News," *Not Even Wrong* (blog), August 25, 2012, www.math.columbia.edu/~woit/wordpress/
6. Penrose, 928.
7. B. Schroer, arXiv.org/abs/0905.4006, 40, see also arXiv.org/abs/physics/0603112.
8. S. Glashow, in *The Superworld I*, ed. A. Zichichi (New York: Plenum, 1990), 250.
9. Smolin, 280.
10. B. Schroer, arXiv.org/abs/0905.4006, 52.
11. Davies and Brown, chap.9, 192ff.
12. L. Krauss, *Quantum Man* (New York: W. W. Norton, 2011), 254.
13. Greene, 121.
14. Horgan,.67.
15. Horgan, 65.
16. Howard Georgi, in *The New Physics*, ed. Paul Davies (Cambridge: Cambridge University Press 1989), 446.
17. G. Kane, *Nature* 480 (December 16, 2011), www.nature.com/news/particle-physics-is-at-aturning-point-1.9675.
18. F. Wilczek, arXiv.org/abs/0708.4361, 19.
19. An example is Paul Davies and John Gribbin, *The Matter Myth* (New York: Simon and Schuster, 2007).
20. E. Witten. "Reflections on the fate of spacetime." *Physics Today*, April 1996, 24.
21. Julian Barbour, "The Third Culture," www.edge.org/documents/archive/edge60.html.
22. E. Witten, *Nature* 438 (2005): 1085.
23. Albrecht Fölsing, *Albert Einstein*. (Berlin: Suhrkamp, 1999), 628.
24. P. Ramond, arXiv.org/abs/0708.3656.
25. For example, Lopez-Corredoira (*Against the Tide*), chap. 2.
26. "Witten M-Theory," YouTube, www.youtube.com/watch?v=iLZKqGbNfck, www.youtube.com/watch?v=6PVjNlXj2WQ

CHAPTER 16

1. "The multiverse in three parts: Brian Greene at TED2012," TED blog, blog.ted.com/2012/02/28/the-multiverse-in-three-parts-brian-greene-at-ted2012/.
2. *Spektrum der Wissenschaft* 2009/05, 39.
3. Lisa Randall, "Theories of the Brane," www.edge.org/3rd_culture/randall03/randall03_print.html.
4. "Raven paradox," *Wikipedia, the Free Encyclopedia*, en.wikipedia.org/wiki/Raven_paradox.
5. Taleb, 39.
6. Gleick, 391.
7. Cornell (ed.), chap. 3, 72.
8. Lindley (EoP), 106.
9. Dario Antiseri, *Popper's Vienna 1870–1930* (Aurora, CO: The Davies Group, 2006), 12.

CHAPTER 17

1. Gleick, 283.
2. J. P. Ralston, arXiv.org/abs/1006.5255.
3. R. Bernabei et al., *Eur. Phys. J.* C56 (2008): 333–355, arXiv:0804.2741.

4. Y. Tomozawa, arXiv.org/abs/0806.1501.
5. M. Baldi et al., *Mon. Not. Roy. Astron. Soc.* 412 (2011): L1, arXiv.org:1006.3761.
6. See "Cowan–Reines neutrino experiment," *Wikipedia, The Free Encyclopedia,* en.wikipedia.org/wiki/Cowan-Reines_neutrino_experiment.
7. Feynman (QED), chap. 3.
8. Rith et al., "The mystery of the nucleon spin," *Scientific American,* July 1999, 58.
9. Lindley (EoP), 191
10. Lindley (EoP), 190.
11. Hu et al., *Nature* 386 (1997): 37.

CHAPTER 18

1. Rosenthal-Schneider, 27.
2. D. Sciama, *Mon. Not. Roy. Astron. Soc.* 113 (1953): 34.
3. Feynman (Lectures II), chap. 28; Landau, § 75.
4. C.F. Frank, *Proc .Roy. Soc. Lond. A* 62 (1949):131.
5. M. Beller, *Physics Today* 51 (1998): 29–34.
6. G. Lochak. "De Broglies initial concept of de Broglie waves," S. Diner (Ed.), *The Wave-Particle Dualism,* (Netherlands: Springer, 1983), 1 ff.
7. P. A. M. Dirac, *Proc. Roy. Soc. Lond. A* 165 (1938): 199.
8. See also Steven Weinberg, *Gravitation and Cosmology* (New York: Wiley, 1972), eqn. 16.4.6.
9. See "Variable speed of light," *Wikipedia, The Free Encyclopedia,* http://en.wikipedia.org/wiki/Variable_speed_of_light.
10. J. Broekaert, *Found.Phys.*38 (2008): 409–435, arXiv:gr-qc/040501, ref.[70].
11. R. Dicke, *Rev. Mod. Phys.* 29 (1957): 363; see also A. Unzicker, *Ann.Phys.* 18 (2009): 53–70, arXiv:0708.3518.
12. A. Einstein, *Ann. Phys.* 38 (1912): 355.
13. Rosenthal-Schneider, 24, 27.

CHAPTER 19

1. See Tony Rothman's interesting comment in "Th e Man behind the Curtain," *American Scientist* 99, no. 3 (May-June 2011): 3, www.americanscientist.org/issues/pub/the-man-behind-the-curtain/3.
2. A. Einstein, *Ann.Phys.* 102 (1930): 685–697.
3. B. A. Bassett et al., *Rev. Mod. Phys.*78 (2006) : 537–589, arXiv:astro-ph/0507632.
4. B. G. Charlton, *Medical Hypotheses* 71(2008):327–329.
5. "Albert Einstein," *Wikiquote,* de.wikiquote.org/wiki/Albert_Einstein.
6. Smolin, 218.
7. Smolin, xi.
8. Einstein, chap.V 7, --about the development of the theory of general relativity.
9. Penrose, 1026.
10. Gleick, 232.
11. Smolin, xxii.
12. Smolin, xii.

CHAPTER 20

1. Interview, *Die Welt,* January 29, 1990, www.neundorf.de/Kritik/kritik.html,
2. Horgan, 65.
3. Karl Popper, *Ich weiß daß ich nichts weiß* (Berlin: Ullstein, 1991).
4. "Peer Review," *Wikipedia, The Free Encyclopedia,* en.wikipedia.org/wiki/Peer_review.
5. See "Ridiculed Discoverers, Vindicated Mavericks," amasci.com/weird/vindac.html
6. K. Krogh, *Class. Quantum Grav.* 24 (2007): 5709–5715.

LITERATURE

The numbers in parenthesis indicate the chapters to which the respective book provides background information.

Arp, Halton. *Seeing Red*. Montreal, QC: Apeiron, 1998. (7)

Baggott, Jim. *Farewell to Reality*. Berkeley, CA: Pegasus Books, 2013. (14, 15)

Barbour, Julian. *The Discovery of Dynamics*. Oxford: Oxford University Press, 2001. (4)

———. *The End of Time*. Oxford: Oxford University Press, 2000. (4)

Bell, John. *Speakable and Unspeakable in Quantum Mechanics*. Cambridge: Cambridge University Press, 1988. (11)

Byers, William. *The Blind Spot: Science and the Crisis of Uncertainty*. Princeton, NJ: Princeton University Press, 2011.

Collins, Harry. *Gravity's Shadow*. Chicago: University of Chicago Press, 2004. (19, 20)

Cornell, James, ed. *Bubbles, Voids and Bumps in Time*. Cambridge and New York: Cambridge University Press, 1992. (chap. 6, 7, 10)

Davies, P. C. W., and Julian Brown. *Superstrings*. Cambridge: Cambridge University Press, 1992. (15)

Einstein, Albert. *The World As I See It*. New York: Open Road, 2011. (4)

Feldman, Burton. *The Nobel Prize: A History of Genius, Controversy and Prestige*. New York: Arcade Publishing, 2000. (16, 20)

Feyerabend, Paul. *Against Method*. London and New York: Verso, 1975. (16)

Feynman, Richard. *Lectures*. Vol. II. New York: Basic Books, 2011. Chap. 27+28. (4).

———. *QED*. Princeton, NJ: Princeton University Press, 1988. (12, 18)

Fuller, Steve. *Kuhn vs. Popper*. New York: Columbia University Press, 2005. (16)

Galison, Peter. *How Experiments End*. Chicago: University of Chicago Press, 1987. (20)

Gleick, James. *Genius*. New York: Pantheon Books, 1992. (12)

Greene, Brian. *The Elegant Universe*. New York: Vintage, 2000. (1, 15)

Guth, Alan. *The Inflationary Universe*. New York: Basic Books, 1998. (10)

Hawking, Stephen. *A Brief History of Time*. New York: Bantam, 1998. (11).

———. *A Briefer History of Time*. New York: Bantam, 2008. (11)

Heisenberg, Werner. *Physics and Beyond*. New York: Harper & Row, 1972. (11, 12)

Horgan, John. *The End of Science*. Reading, MA: Addison-Wesley, 1996. (10, 14, 15)

Jones, Sheilla. *The Quantum Ten*. Oxford: Oxford University Press, 2008. (11)

Jungk, Robert. *Brighter Than a Thousand Suns*. Mariner Books, 1970. (20)

Krag, Helge. *Higher Speculations*. Oxford: Oxford University Press, 2011.

Kuhn, Thomas. *The Structure of Scientific Revolutions*. Chicago: University of Chicago Press, 1996. (16, 17)

Landau, Lev. *Theoretical Physics* II. Oxford: Pergamon Press, 1975. (4)

Laughlin, Robert B. *A Different Universe*. New York: Basic Books, 2006. (15, 20)

Lederman, Leon. *The God Particle*. Boston, MA: Houghton Mifflin, 2006. (12, 13)

Lerner, Eric. *The Big Bang Never Happened*. New York: Vintage, 1992. (7)

Lindley, David. *The End of Physics*. New York: Basic Books, 1994. (1, 14, 20).
——. *Uncertainty*. New York: Anchor, 2008. (11, 14)
López Corredoira, Martín. *Against the Tide*. Boca Raton, FL: Universal Publishers, 2008. (3, 20).
——. *The Twilight of the Scientific Age*. Brown Walker Press, 2013. (3, 20)
Magueijo, João. *Faster Than Light*. New York: Penguin Books, 2003. (10)
Parker, Barry. *The Vindication of the Big Bang*. New York: Plenum Press, 1993. (7)
Penrose, Roger. *The Road to Reality*. New York: Vintage, 2004. (10, 16)
Pickering, Andrew. *Constructing Quarks*. Chicago: University of Chicago Press, 1984. (12, 13)
Popper, Karl. *The Logic of Scientific Discovery*. London and New York: Routledge, 2002. (16)
Rosenthal-Schneider, Ilse. *Begegnungen mit Einstein, von Laue und Planck*. Braunschweig: Vieweg, 1988.
Sagan, Carl. *Cosmos*. New York: Ballantine Books, 1985.
Sanders, Robert. *The Dark Matter Problem*. Cambridge: Cambridge University Press, 2010. (6, 7)
Schrödinger, Erwin. *My View of the World*. Cambridge: Cambridge University Press, 2008. (11, 20)
Segrè, Emilio. *From X-Rays to Quarks*. Berkeley: University of California Press, 1980. (13, 20)
Singh, Simon. *Big Bang*. New York: Harper, 2005. (1, 3)
Smolin, Lee. *The Trouble with Physics*. Boston, MA: Houghton Mifflin, 2006. (1, 15, 20)
Taleb, Nicholas Nassim. *The Black Swan*. New York: Random House, 2007. (13, 16, 20)
Taubes, Gary. *Nobel Dreams*. New York: Random House, 1987. (12, 13, 20)
Veltman, Martinus. *Facts and Mysteries in Elementary Particle Physics*. River Edge, NJ: World Scientific, 2003. (12, 13)
Weinberg, Alvin. *Reflections on Big Science*. Cambridge: MIT Press, 1969. (13, 20)
Weinberg, Steven. *The First Three Minutes*. New York: Basic Books 1993. (10)
Whittaker, Sir Edmund. *The History of the Theories of Aether and Electricity*. New York: Dover, 1951. (4)
Will, Clifford. *Was Einstein Right?* New York: Basic Books, 1989. (4, 10)
Woit, Peter. *Not Even Wrong*. New York: Vintage, 2006. (1, 14, 15)

INDEX

Acatama Large Millimeter Arra (ALMA), 26
Aguirre, Anthony, 9, 76
Andromeda galaxy, 31, 33–4, 86
anthropic principle, 191–2, 201, 214
antimatter, 103, 206, 209
Apollo space missions, 23, 67
Aristarchus of Samos, 200
Arp, Halton, 91–2, 248
ArXiv.org, 185–6, 244–5
Ashtekar, Abhay, 132
atomic clocks, 22, 47, 213

Baade, Walter, 27, 34–5
Baez, John, 139
Balmer, Johann Jakob, 114–15, 222
Barbour, Julian, 48–9, 65, 184, 234
Barrow, John, 50, 128
Bartelmann, Matthias, 93
baryonic acoustic oscillations, 99–100
Berkeley, George, 164
Berry, Michael, 223
Bessel, Friedrich Wilhelm, 32
Big Bang, 6, 15, 24–5, 50, 91, 104–5, 123–4, 129,
    131, 135–9, 211
big science, 157, 200, 203, 234–6, 239–49
binary stars, 55–6
black holes, 63–4, 73, 131, 135, 137–8, 171, 193,
    219–20, 231
Bogdanov Affair, 244
Bohr, Niels, 16, 115, 133–4, 146, 182, 222–3,
    234, 236
Born, Max, 102, 134, 169, 223
Bose, Satyendranath, 161
Bothun, Gregory, 81
Brahe, Tycho, 27
branes, 93, 165, 171, 173, 214
Bullet Cluster, 74–5

Cartan, Ellie, 52
Cavendish, Henry, 59–61

Cepheid stars, 33–6, 38
CERN, 15, 26, 137–8, 152, 154, 158–9, 181
charged coupled devices (CCDs), 20–1
COBE satellite. See NASA Cosmic Background
    Explorer (COBE) satellite
CODATA, 60–1
cold dark matter (CDM), 81, 90–2
concordance model of the universe, 9–10, 45,
    90, 119
consensus cosmology, 41
constants of nature, 11, 60–3, 67, 156, 163–4,
    170, 191–2, 217–9, 222, 227–8
Copenhagen Interpretation, 223
Copernican Revolution, 20, 204
Copernicus, Nicolaus, 94, 200
cosmic microwave background, 3–4, 25, 41, 82,
    95–107, 114, 122–4, 128–31, 136, 207
Cosmological Evolution Survey (COSMOS),
    28–9

dark energy, 7–9, 39–42, 84–90, 94, 98, 118, 171,
    198, 207, 210
dark matter, 14, 42, 45, 70–9, 112–14, 117–18
    cold dark matter (CDM), 81, 90–1
    dark energy vs., 84, 86
    "discovery" of, 5, 70
    fluctuations of, 97
    halo, 70, 72–6, 83
    predictions of, 74–5
    rotation curves and, 82, 71
    wave harmonics and, 98
Dawkins, Richard, 167–8
de Broglie, Louis-Victor, 142, 222, 224, 234, 245
de Haas, W. J., 143–4, 247
de Vaucouleurs, Gerard, 87, 91
Dicke, Robert, 25, 49–50, 121, 127, 220n, 227
Dirac, Paul, 12, 16, 62, 66–7, 163–5, 167–8, 211,
    218, 224–6, 228, 233, 237, 244
Disney, Mike, 40, 82–3, 90, 107, 113–14, 119,
    168n

distance ladder, 32–3
Doppler shift, 5, 37, 69–70, 83, 96
Durrer, Ruth, 76, 88, 166
dwarf galaxies, 73, 82, 89

Eddington, Arthur, 12, 54, 64–6, 226
Einstein, Albert, 29, 51, 54, 68, 76, 85–6, 127, 133, 139, 143–4, 169, 231–3
    on beauty, 233
    Bose and, 161
    Cartan and, 52
    $E = mc^2$ of, 46, 53, 103, 135, 208, 218
    Mach and, 66, 219
    notebooks of, 52–3
    photoelectric effect, 177
    on simplicity, 11
    on theories, 217
    on working alone, 8, 234–5
    See also theory of general relativity; theory of special relativity
Einstein Papers Project, 52
Eisenstein, Daniel, 99
electrodynamics, 20, 51n, 52–4, 125, 138, 145–6, 162, 163, 217–22, 230, 236
epicycles, Ptolemaic, 11, 39, 42, 93–4, 157, 195, 197, 212, 233, 237, 249
equivalence principle, 47, 53, 77
European Extremely Large Telescope (E-ELT), 26
European Space Agency (ESA), 22–3, 27, 32, 60, 102
Everett, Hugh, 170

Faraday, Michael, 20, 200
Fermi, Enrico, 161, 205, 209, 212
Fermilab, 14n, 126
fermion, 161–2
Feyerabend, Paul, 193
Feynman, Richard, 12, 16, 152, 155, 196, 199n, 210, 247
    on calculation of radiation, 53
    and current theoretical physics, 145–7, 162–3, 214, 223
    on string theory, 13, 177
fine structure constant 124, 163, 218, 254
Fischer, Bobby, 248
flatness, 65, 121–2, 128, 211
Fraunhofer, Joseph, 104
free parameters 10–1, 15, 112, 127–8, 158, 195, 204, 207, 209, 215–28, 233–4, 237, 252
Frenk, Carlos, 75

GAIA spacecraft, 32
Galileo Galilei:
    discoveries of, 19–21, 45–6, 60, 68–9, 96
    telescopes of, 19–20, 27, 29, 69
Geller, Margaret, 87, 199

Gell-Man, Murray, 127, 147–8, 198, 247
general relativity, theory of, 7, 12, 22–3, 45–6, 50–5, 57, 59, 76, 122, 124, 134n, 213, 219–20, 227, 229–30
Georgi, Howard, 145, 181
Ginsparg, Paul, 185–6
Glashow, Sheldon, 127, 144, 176–7, 187, 241
globular clusters, 36–8, 56, 72–3, 78, 117, 231
Grand Unified Theories (GUTs), 123, 145
gravitational constant G, 11, 15, 23, 47, 59–61, 64–8, 79, 132–3, 156, 164, 195, 219–20, 225, 227
    See also Newton, Isaac: law of universal gravitation
Gravity Recovery and Climate Experiment (GRACE), 22
Greene, Brian, 4, 13, 173, 177–9, 231
Gross, David, 175, 179, 186–7
group theory, 145
Grundlach, Jens, 61
Guth, Alan, 4, 123, 126–8, 195, 235

Hahn, Otto, 212
Hawking, Stephen, 16, 132, 134–5, 138–9, 158, 167, 170, 200
Heisenberg, Werner, 16, 134–7, 146, 148, 153, 156n, 174, 223, 234–6
Heisenberg's uncertainty principle, 135, 168
Heuer, Rolf-Dieter, 154
Higgs boson, 137, 197, 199n
HIgh Precision PARallax COllecting Satellite (HIPPARCOS), 32
Hilbert, David, 102, 141, 229–30, 236
Hooper, Dan, 82, 117, 145
Horgan, John, 119, 155–6, 171, 178–80
horizon problem, 122
Hoyle, Fred, 91, 131, 191, 214
Hubble, Edwin, 5–6, 31–3, 64, 87–90, 96, 163, 207, 234
Hubble constant, 32, 34–6, 38, 40, 88, 94
Hubble Space Telescope, 20, 26–9, 34
Huchra, John, 87

inertia, 47, 53, 55, 65, 152, 220
inflation theory, 15–16, 93, 121–9, 131, 138, 166, 168, 170–1, 195, 199, 205, 211, 213, 231–2, 234, 237, 240, 249, 252
ironic science, 171

James Webb Space Telescope (JWST), 26–7
Jansky, Karl, 24–5
Josephson, Brian, 245

Kaku, Michio, 137, 178
Kane, Gordon, 181
Kant, Immanuel, 31, 118, 194
Kepler, Johannes, 4, 27, 106

Kepler's law of planetary motion, 71
Krauss, Lawrence, 40–1, 165, 168
Kuhn, Thomas, 94, 184, 193, 203–4

Lambda (cosmological constant), 84–5206-7
Lambda CDM model, 81, 90–3
Landau, Lev, 7–8, 34–5, 41, 152, 220
Large Hadron Collider (LHC), 137–8, 152, 155, 159, 181, 199n, 235
Large Number Hypothesis, 164, 254
Large Synoptic Survey Telescope (LSST), 27
Laser GEOdynamic Satellite (LAGEOS), 23
Laser Interferometer Space Antenna (LISA), 22
Laughlin, Robert B., 171, 248
Leavitt, Henrietta, 33
Leibundgut, Bruno, 41
Linde, Andrej, 166, 171
Lindemann, Ferdinand, 137, 156n
Lindley, David, 157, 159, 200, 210–11, 237n
Loop Quantum Gravity, 132, 139
López Corredoira, Martín, 92, 117
Lorentz, Hendrik, 54
low surface brightness (LSB) galaxies, 81, 90
lucky imaging, 26
Lunar Laser Ranging, 23, 67
Lyman-alpha line, 115–16
Lyman-alpha forest, 116, 118

Maartens, Roy, 88–9
Mach, Ernst, 49–50, 65–6, 171, 219–20, 225, 227–8
Mach's principle 49–50, 65, 225
Magueijo, João, 117, 125, 127, 187
Malmquist bias, 34
Massive Compact Halo Objects (Machos), 73
Maxwell, James, 20, 54, 125, 222, 230, 232
Michell, John, 63
Michelson, Albert, 200
Milgrom, Mordehai, 77
Milky Way galaxy, 24–5, 28n, 31–3, 37, 64, 70–5, 83, 86–7, 96, 100, 102, 106, 108, 112, 206
Millennium Simulation, 111–12
Milner, Yuri, 240
MOdified Newtonian Dynamics (MOND), 76–9, 207
Mount Wilson Observatory, 26, 31, 34
multiverses, 16, 170–1, 174, 186, 252

NASA Cosmic Background Explorer (COBE) satellite, 3–4, 25, 96–7, 101–2, 107
neutrinos, 98, 106–8, 113, 157, 207–8, 247
Newton, Isaac:
    law of universal gravitation, 7–8, 10–11, 20, 45–6, 49, 59, 61–3, 68, 76–8, 195, 219–20
    laws of motion, 45, 57, 221
    See also gravitational constant G
Noddack, Ida, 212

Noether, Emmy, 141–3

Occam's razor, 195
octupole anomaly, 100–2
Ostriker, Jeremiah, 70, 113

PAMELA satellite, 206
parallax measurement, 32–3
Pauli, Wolfgang, 145, 148, 167, 174–5, 198, 208–9, 223
Peebles, James, 70, 89
peer review, 243, 245
Penrose, Roger, 105, 123–4, 129, 133, 148, 175, 184, 199, 235
Penzias, Arno, 25
Perlmutter, Saul, 38
philosophy of science, 193–4
phlogiston theory, 75–6
Pickering, Andrew, 148, 209, 246
Pierre Auger Observatory, 26
Pioneer spacecraft, 8, 77n
Planck, Max, 3, 25n, 37, 63, 85
Planck scale (Planck length), 15, 80, 131–5, 169, 196, 219, 225–6
Planck spacecraft, 81, 96, 101–2, 128
Planck's constant, 63, 85, 143, 156, 218, 222, 224
Planck's radiation law, 3, 37, 72, 222
Popper, Karl, 126, 184, 193–4, 196–200, 203–4, 239–41, 248
precision cosmology, 98, 102
Ptolemy, 11, 39, 157, 184, 195, 204.
    See also epicycles, Ptolemaic
pulsars, 22, 56, 166

quantum electrodynamics, 53, 145–6, 217–20, 236
quantum field theory, 12, 16, 146, 218, 236
quantum foam, 16, 169
quantum mechanics, 11–12, 16, 20, 66, 85, 142, 157, 164, 169–70, 213, 222–4, 229–30, 234, 237
quarks, 147–9, 153, 155n, 157–8, 198, 209–10
quasars, 20–1, 24–5, 28, 30, 92, 103–4, 105n, 112, 115–16
Quinn, Terry, 61
quintessence, 9, 14, 84–5

Rabi, Isidoor Isaac, 209
Ralston, John, 205–6
Randall, Lisa, 138, 165, 167, 185, 194, 242
redshift, 5–6, 26, 31, 40, 54, 56, 64, 85, 87, 95, 103, 114–16
Rees, Martin, 11, 124–5, 166
reionization, 117, 211
relativity. See theory of general relativity; theory of special relativity
Riess, Adam, 38

Robitaille, Pierre-Marie, 100
rotation curves of galaxies, 71–2, 77, 79, 82
Rubin, Vera, 70–1
Russell, Bertrand, 129, 241
Rutherford, Ernest, 158

Salucci, Paolo, 82
Sandage, Alan, 35–6, 87, 94
Sanders, Robert, 70, 72, 74, 112
Schild, Rudy, 20–1, 35, 41
Schmidt, Brian, 38
Schrödinger, Erwin, 66, 97, 146, 157–8, 169–70, 213, 222, 230, 234, 236
Schroer, Bert, 175–6
Schwarzschild radius, 63–4, 137n
Sciama, Dennis, 65, 219–20, 225
Search of Extraterrestrial Life (SETI), 70
Shapiro, Irwin I., 55
Shelton, Ian, 106
Shostak, Seth, 70
Sloan Digital Sky Survey (SDSS), 28, 99, 249
Smolin, Lee, 11, 13–14, 78, 155, 166, 174, 176–7, 183, 186, 204, 233–4, 236
Sokal, Alan, 243–4
special relativity, theory of, 45–7, 55, 66, 127, 182, 221, 224, 226–7
spin crisis, 210
standard candles, 33, 39
standard model of cosmology, 15, 41, 67, 78, 81, 84, 89–94, 98, 101, 117–8, 129, 213, 225–6, 234–5, 249, 252
standard model of particle physics, x, 10, 14, 141, 149–61, 168, 182, 187, 195, 197, 210, 213, 231, 234 237, 252–3
Steinhardt, Paul, 128
Strassmann, Fritz, 212
string theory, 13–15, 86, 93, 133, 136–7, 166, 186, 173–87, 191–6, 204
supernovae, 6, 9, 22, 28, 30, 36–41, 56, 84–5, 106, 112, 114
superstring theory, 12–13, 84, 175–7, 182
supersymmetry (SUSY), 16, 74, 159–62, 174, 182, 210–11, 237, 242, 249
Susskind, Leonard, 84, 126, 187, 192, 196
Sylos Labini, Francesco, 89, 99, 117–18
symmetry, 125, 141–5, 167, 194, 233–4
    See also supersymmetry (SUSY)
symmetry breaking, 211, 214

't Hooft, Gerardus, 174–5
Taleb, Nassim, 153, 158, 196, 252
Taylor-Hulse pulsar, 56
telescopes:
    Chandra, 26
    European Extremely Large Telescope, 26
    Fermi, 26, 75
    of Galileo, 19–20, 27, 29, 69
    Hubble, 20, 26–9, 34
    James Webb Space Telescope, 26–7
    Large Synoptic Survey Telescope, 27
    SWIFT, 26
    satellite, 4–5, 19–29, 72
Theories of Everything (TOEs), 134, 173–5, 200, 242, 252
theory of general relativity, 7, 12, 22–3, 45–6, 50–5, 57, 59, 76, 122, 124, 134n, 213, 219–20, 227, 229–30
theory of special relativity, 45–7, 55, 66, 127, 182, 221, 224, 226–7
time dilation, 46–7

universe:
    age of, 22, 32, 35, 37–8
    concordance model of, 9–10, 45, 90, 119
    voids in, 87, 89, 116, 211

variable speed of light, 127, 227
Very Long Baseline Interferometry (VLBI), 24

wave-particle duality, 142, 205
weakly interacting particles (WIMPs), 75
Wegener, Alfred, 241–2
Weinberg, Steven, 16, 105, 127, 144, 166
Wheeler, John, 54, 169–70
White, Simon, 78–9
Wilczek, Frank, 11, 113, 182, 225n
Wilkinson Microwave Anisotropy Probe (WMAP), 25, 81, 96, 100–4, 107, 128
Wilson, Robert, 25
Wiltshire, David, 88–9
Witten, Edward, 175–6, 179–80, 182–7
Woit, Peter, 159–60, 174–5, 179, 183, 186, 196

Zwicky, Fritz, 5–6, 72